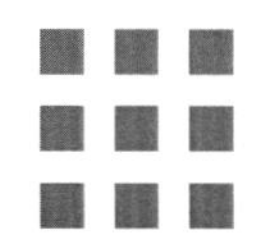

Prevention of Valve Fugitive Emissions in the Oil and Gas Industry

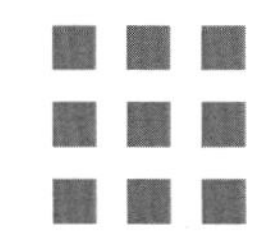

Prevention of Valve Fugitive Emissions in the Oil and Gas Industry

Karan Sotoodeh

Gulf Professional Publishing is an imprint of Elsevier
50 Hampshire Street, 5th Floor, Cambridge, MA 02139, United States
The Boulevard, Langford Lane, Kidlington, Oxford, OX5 1GB, United Kingdom

Library of Congress Cataloging-in-Publication Data
A catalog record for this book is available from the Library of Congress

British Library Cataloguing-in-Publication Data
A catalogue record for this book is available from the British Library

ISBN: 978-0-323-91862-6

For information on all Gulf Professional Publishing publications
visit our website at https://www.elsevier.com/books-and-journals

Publisher: Joe Hayton
Senior Acquisitions Editor: Katie Hammon
Editorial Project Manager: Chris Hockaday
Production Project Manager: Sojan P. Pazhayattil
Cover Designer: Alan Studholme

Typeset by TNQ Technologies

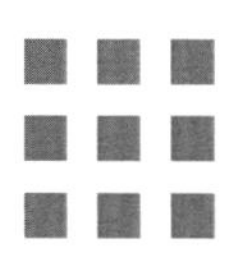

Contents

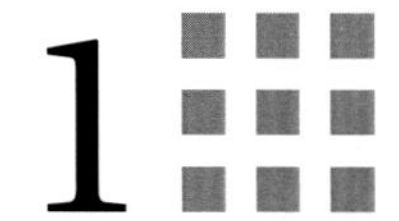

1 Terms and definitions

A

Acid rain: Acid rain or acid deposition is a broad term that refers to acid components such as sulfuric or nitric acid that fall to the ground by mixing with rain, snow, fog, or dust. Acid rain is an environment pollutant that has harmful effects on plants, animals, and infrastructure.

Actuator: A mechanical or electrical device or component installed on a valve to automatically operate it. Actuators typically work with electricity, hydraulic fluid, or air. The power of electricity or hydraulic fluid or pneumatic is transferred into mechanical force to operate the valve.

Advanced oil recovery: The oil reservoir typically has enough pressure to move the petroleum from the subsurface to the surface where the production equipment is located. However, the reservoir pressure declines over time and oil production is reduced as a result. Different approaches, such as injecting water or gas into the reservoir or installing a pump in the well, enhance oil recovery from the reservoir; these processes are called "advanced oil recovery." Advanced oil recovery accomplished by injecting gas into the reservoir is called "gas lift."

Ambient temperature: A temperature that is between 15°C (59°F) and 40°C (104°F), as per API standards 622 and 624 regarding valve packing fugitive emission tests.

Anti-explosive decompression (AED): Explosive decompression is a type of failure for O-rings used in high pressure applications and gas services. Explosive decompression is also called rapid gas decompression (RGD) in which the gas permeates into the sealing due to sudden pressure drop and leads to sealing failure and leakage. Therefore, O-rings should have an AED property under specific conditions such as gas service and high-pressure applications to prevent failure due to RGD. Lip seals are AED by nature.

API: The American Petroleum Institute (API) is the largest United States trade association for the oil and gas industry. Its headquarters are located in Washington, DC. Standards developed by the API are widely used in the oil and gas industry.

Aramid fibers: A class of heat-resistant, strong synthetic fibers. Aramid fibers are used as a material for flat gaskets.

Prevention of Valve Fugitive Emissions in the Oil and Gas Industry. https://doi.org/10.1016/B978-0-323-91862-6.00004-6

ASME: The American Society of Mechanical Engineers (ASME) is an American professional association that promotes the art, science, and practice of multidisciplinary engineering and allied sciences around the globe.

ASTM: The American Society for Testing and Materials (ASTM) is an international organization that develops and publishes technical standards for a wide range of materials and products. ATSM is widely used to define the chemical and mechanical properties, as well as the tests and other requirements for piping and valve metallic components.

Auxiliary valve connections: Different connections, such as plugs or flanges, may be used on the upper and lower areas of the valve body. The purpose of these plugs and flanges is to vent or drain the fluid trapped in the body cavity of the valve to prevent the fluid from accumulating or becoming overpressurized. Plugs or flanges installed on the higher part of a valve are typically used for venting gases, whereas plugs or flanges used on the low areas of a valve body are used for the drainage of liquid services. Plugs are typically threaded into the body of the valve and sealed with Teflon or Loctite. The other purpose of a type of auxiliary connection on a valve body is to provide bypass (see Fig. 1.1) in order to equalize the fluid pressure across the valve so that it can be operated more easily against less pressure differential. Operating the valve against less pressure differential provides smooth operation and requires less torque or force, and reduces the likelihood of wear and tear on the valve internals. Fig. 1.2 illustrates vent and drain plugs on a small ball valve. Fig. 1.3 illustrates vent and drain flanges on a medium size ball valve. As a general rule, plugs on body cavities as auxiliary connections are typically used for small, low-pressure class valves. Flange connections on the body cavity are more robust than plug connections and are common for larger valves in higher pressure classes.

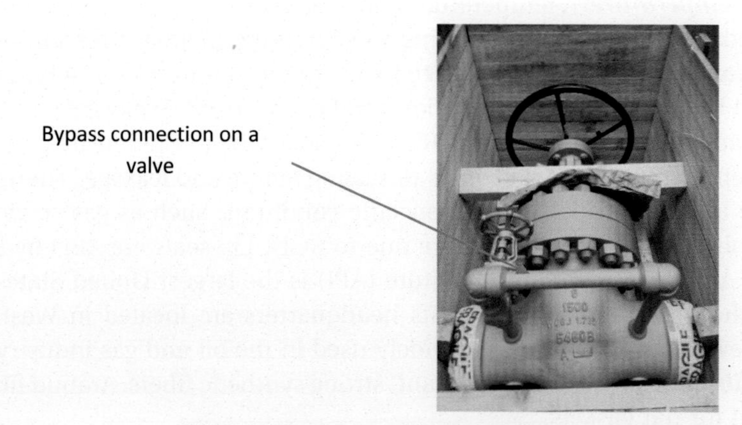

FIGURE 1.1 Bypass connection on a valve.

FIGURE 1.2 Drain and vent plugs on the body cavity of a ball valve.

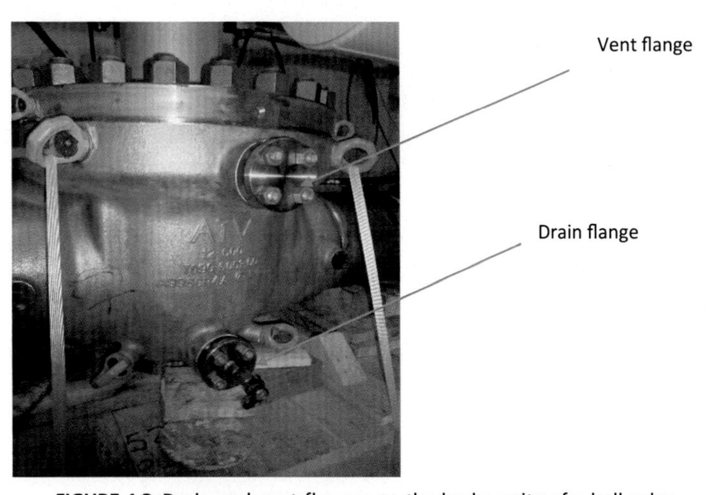

FIGURE 1.3 Drain and vent flanges on the body cavity of a ball valve.

B

Ball valve: A type of quarter-turn valve used to start or stop the fluid (on/off purpose). The shape of the valve closure member is like a ball with a hole inside. When the hole of the ball is placed parallel to the fluid flow, the valve is open and the flow passes through the valve. The rotation of the closure member 90° closes the valve, as the hole in the ball stands perpendicular to the flow direction and the solid part of the closure member stops the fluid.

Bar: A metric unit of pressure, but not part of the International System of Units (SI). It is defined as exactly equal to 100,000 Pascal (Pa) (100 kPa), or slightly less than the current average pressure at sea level (approximately 1.013 bar). Each bar is approximately 14.7 times of 1 psi (pound per square inch).

Bellows: A type of metallic stem seal for valves that can provide very tight sealing and reduce fugitive emission significantly in comparison with other stem seals like graphite packing, PTFE, or elastomeric O-rings. It is in the shape of a bellow and is typically welded to the stem and bonnet of the valve. Fig. 1.4 illustrates stem seal bellows for a globe valve.

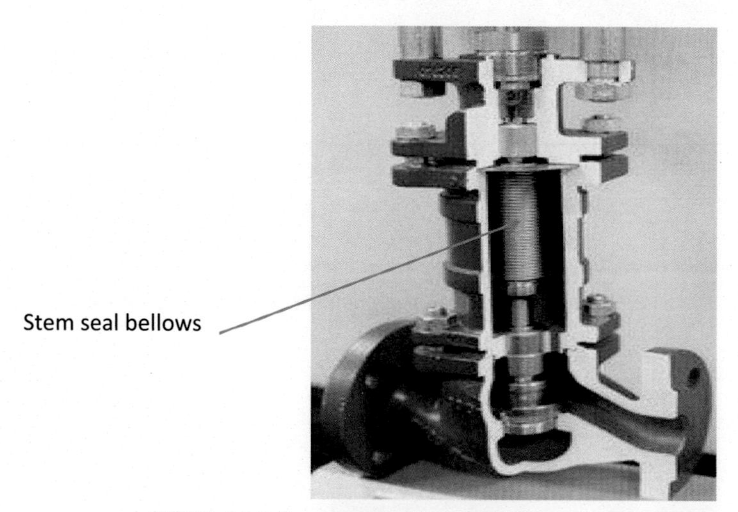

FIGURE 1.4 Bellows stem seal in a globe valve.

Bevel end: A bevel end, also called a bevel edge, is an end or edge of a pipe that is not perpendicular to the face of the pipe. A bevel is made on the end of the pipe through a pipe beveling process. Beveling is performed to prepare the pipe ending for welding. Other piping components such as fittings and flanges may have beveled ends. Fig. 1.5 illustrates a bevel ending of a pipe connection, such as piping or fitting. The bevel end is prepared for a buttweld connection.

FIGURE 1.5 Bevel end preparation.

Body: The casing part of a valve that is under pressure from the fluid service inside the valve. The body of a valve is categorized as a pressure-containing part, meaning that its failure to function leads to the leakage of fluid from inside the valve to the environment.

Body seals: Seals that are used between the body pieces of valves. Body seals are explained in detail in Chapter 2. Stem seals are not included into this category of seals. Different types of gaskets as well as O-rings can be used as body seals.

Bonnet: A pressure-containing valve component placed on the top of the valve body; the stem of the valve passes through it. The bonnet could be welded to the body or attached to the body with bolting with a gasket between these two joints to prevent leakage.

Bottom cover: A bottom cover, also called a bottom flange, is installed at the bottom of the valve body on the opposite side of the valve from the bonnet.

Bushing: See stem nut.

Butterfly valve: A quarter-turn valve like a ball valve used for both flow regulation and fluid isolation. The closure member of the valve is a disk that rotates 90° between open and closed positions. Butterfly valves can be used instead of ball valves as a less costly choice for fluid isolation in utility services such as water.

Buttweld connection or joint: Buttweld end connections are prepared by beveling each end of a valve or pipe as per an international standard, such as ASME B16.25, "Buttwelding ends," or based on narrow gap welding, which is a nonstandard bevel preparation. Buttweld connection after applying 100% radiography test has joint efficiency of 100% with zero leakage possibility. Fig. 1.6 illustrates end preparation based on the ASME B16.25 standard. Fig. 1.7 illustrtaes an automatic buttweld connection between two piping joints.

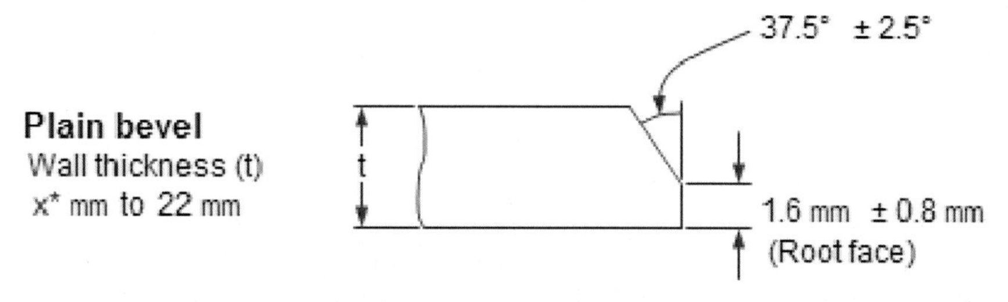

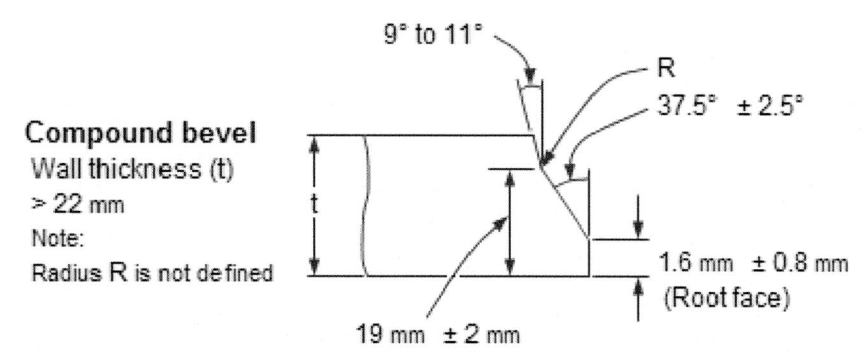

FIGURE 1.6 Buttweld end preparation as per ASME B16.25. *Courtesy: Elsevier.*

FIGURE 1.7 Buttweld connection schematic. *Courtesy: Shutterstock.*

C

Cavitation: Cavitation is a phenomenon in which a rapid drop in pressure drop below the liquid vapor pressure in a liquid service leads to the formation of bubbles that burst firmly and make pits or irregularities inside the valve. Valves that create high pressure drop, such as globe valves in liquid services, are exposed to cavitation. Fig. 1.8 illustrates cavitation in the plug of a globe valve. Therefore, the API 623 standard was developed in 2013 to design a more robust globe valve with much less risk of cavitation.

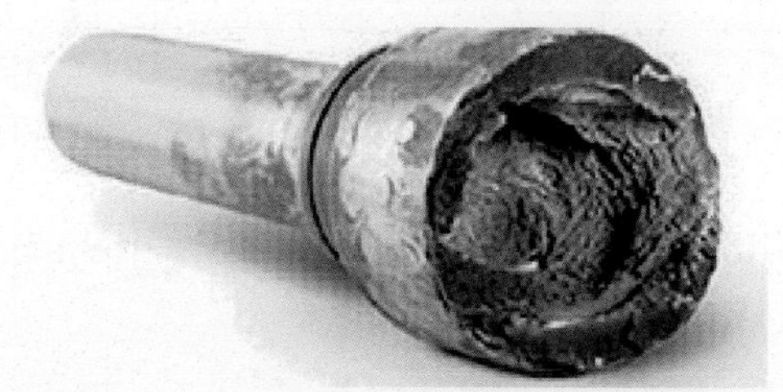

FIGURE 1.8 Cavitation in the plug of a globe valve. *Courtesy: Valve World.*

Cavity: Some valve designs, e.g., designs for ball valves, cause a body cavity. A cavity is an area that is formed between the valve closure member, such as the ball, and the body of the valve. Fig. 1.9 illustrates the cavity area in red.

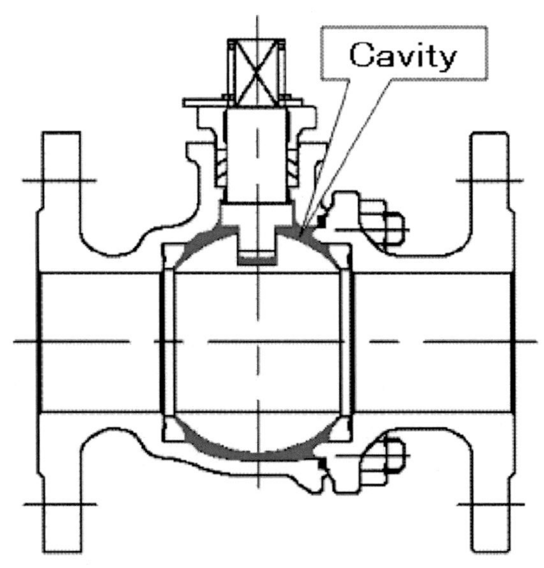

FIGURE 1.9 Cavity area in a ball valve highlighted in red.

Circumferential weld: A type of weld used to join two rounded objects, such as pipe, around their circumference. Conversely, longitudinal weld is a straight weld on the surface of the piping. Fig. 1.10 shows both circumferential and longitudinal welds on piping connections.

FIGURE 1.10 Circumferential and longitudinal welds. *Courtesy: Shutterstock.*

Closure member: Also called a valve obturator, a closure member is a part of a valve positioned inside the flow path to permit or prevent the flow. Different valves have diverse shapes and types of obturators, such as a ball, gate, disk, or wedge.

Compression packing: Compression packing refers to the soft materials placed in an area around the valve stem, known as the "stuffing box"; the packing is compressed by a valve part called a "gland or gland follower" to create stem sealing.

Compressor: A facility or piece of equipment used to pressurize and move gas in a piping system. Compressors typically have different types of design such as rotary and reciprocating.

Control valve: A type of automated or actuated globe valve used to control or regulate the flow by varying the size of the flow passage. The degree of valve opening and closing is adjusted by a signal from the controller or control room. In general, a control valve is used to control essential variables such as pressure, temperature, flow, and liquid level. In fact, a control valve is the last component in a control loop. The first component in the control loop is a sensor that measures one of the four variables mentioned above. The value measured by the sensor is sent to a controller or logic solver that compares the measured value with what it should be. Based on the evaluation, the controller sends an electricity signal in the form of 4–20 MA (milli-ampere) to adjust the position of the control valve. Fig. 1.11 illustrtaes a control valve as the last element in a plant.

FIGURE 1.11 A control valve in a plant. *Courtesy: Shutterstock*

Corrosion-resistant alloy: A special metallic alloy with excellent properties of corrosion resistance and strength to withstand high pressure and temperature. Typically, nickel alloys and stainless steels such as duplex are considered corrosion-resistant alloys.

Crevice corrosion: This type of corrosion refers to a localized attack on a metal surface that has a crevice or gap inside, e.g., grooves where gaskets or seals are placed. Crevice corrosion on machined grooves where seals sit is very common in nonexotic (noncorrosion resistant) materials such as carbon steel. Crevice corrosion can initiate other types of corrosion such as pitting.

Cycling: Valve cycling means the action of opening and closing the valve. In fact, the valve is opened and closed, or closed and opened in one operation cycle. Opening and closing the valve is more precisely called mechanical cycling, which is performed during the fugitive emission test according to the test standards.

D

Dimension Nominal (DN): DN stands for Dimension Nominal in millimeter to be used to define the size for both piping and valves. DN is calculated by multiplying the pipe or valve size in inches by 25. As an example, a 1″ pipe is equal to DN 25.

Downstream oil and gas: Further processing and purifying of oil and gas that is done in refineries located after the upstream sectors is called downstream. Petrochemical plants are considered to be "downstream" in the oil and gas industry.

Drain: Drain, also called drainage, is defined as the process of releasing unwanted liquid or water to the environment.

Dynamic leak measurement: Measurement of leakage from the valve packing while the valve stem is moving.

E

Elastomer: A type of polymer used for sealing industrial valves characterized by high viscosity and elasticity.

Elgiloy: A corrosion-resistant cobalt alloy with UNS R30003 (UNS is defined further in this chapter). This material is used as a spring material in valves or as a metallic spring inside a lip seal. More precisely, Elgiloy (Co, Cr, Ni) is a cobalt alloy containing 40% cobalt, 20% chromium, 15% nickel, and 7% molybdenum. This material has very good corrosion resistance in general, as well as against seawater.

EPA Method 21: A leak check method established by the United States Environment Protection Agency (EPA) for performing and measuring fugitive emissions from equipment such as valves, flanges, and pumps.

Evaporation: A natural phenomenon in which a liquid turns into a gas or vapor, due to the low pressure of the system as an example. Evaporation can occur in storage tanks that are used in the oil and gas industry to store oil at a low pressure, e.g., atmospheric pressure.

F

Fatigue stress: Failure of a structure or component due to repetition and load cycle. Failure due to fatigue stress can occur due to the frequency of the loads even if the applied stresses are below the allowable stress of the component. Fatigue stress is typically associated with tensile load but it can also occur due to the frequency of compression loads.

Flange: A ring-shaped bulk piping component that is used to join two pieces of pipe together or connect piping to equipment in general. Flanges are typically welded to the connected pipe, but may also be threaded. Two flanges are connected with bolting (bolts and nuts), and a gasket is placed between two flanges for sealing. Fig. 1.12 illustrates a weld neck flange welded to the connected piping. A weld neck flange is a very common type of flange in critical services involving high pressure and temperature and frequent loads.

FIGURE 1.12 A weld neck flange and its welding to a pipe. *Courtesy: Shutterstock*

Flare: A flare, also called as a gas flare, is a system that includes piping and facilities for the transportation and combustion of overpressure and extra gases in petroleum refiners, oil and gas production units, and petrochemical plants. The gas in the flare system is burnt without using the produced heat. Flaring the gas releases a considerable amount of fugitive emissions to the environment.

Fluid service: A general term used to refer to a piping system. A fluid service is designed by considering the application of the piping systems with reference to a combination of parameters such as fluid type (methane, ethane, oil, water, etc.), the state of the fluid (gas, liquid, etc.), and the operating conditions such as pressure and temperature.

Fugitive emission: Fugitive emission is defined as an unintentional and undesirable emission, leakage or discharge of the gases or vapors from pressure containing equipment or facilities, and components inside a plant such as valves, piping flanges, pumps, storage tanks, valves, compressors, etc. Fugitive emission is also known as leak or leakage.

H

HSE: HSE stands for health, safety, and environment. It refers to a methodology that implements practical aspects of protecting the environment and ensuring occupational health and safety. HSE implementation is a concern of companies, especially those in the oil and gas industry, that seek to provide guidelines and procedures from the overarching and general level to the details of design to prevent HSE problems such as oil spillage, fugitive emission, and leakage.

I

Inconel 625: A type of nickel alloy material. The number 625 refers to the grade of the nickel alloy. This exotic and corrosion-resistant material is common in the offshore sector of the oil and gas industry for piping and valves. It has relatively high mechanical strength and can be used in elevated temperatures. Inconel 625 has at least 20% chromium, 8% molybdenum, and more than 50% nickel.

G

Galvanic corrosion: An electrochemical reaction that occurs between metals and materials with different electrical potential and nobility. In the contact between two dissimilar materials, the one that is less Nobel plays the role of an anode and will be corroded in favor of the more Nobel material. Graphite is a very common stem packing material; it is also a very Nobel material that may cause galvanic corrosion of the valve metallic stem in areas that are in contact with it.

Gate valve: An on-off valve that works by inserting a rectangular gate or wedge into the flow of the fluid. Fugitive emission standards for gate valves are covered by API 624 and ISO 15848-1. It should be noted that the stem motion in a gate valve is typically linear, which creates a lot of friction between the valve stem and packing. The friction can cause packing wear and tear as well as leakage. Gate valves are available in different types such as slab, expanding, and wedge gate. A wedge gate valve has a sealing element in the shape of a wedge. A wedge gate is a torque-seated valve, meaning that the wedge expands from both sides due to the stem force and provides sealing. The expansion of the valve closure member due to the stem axial force is called "Wedge effect." An expanding gate valve is another torque-seated valve with a closure member in two sections: one male and the other female. Slab gate valve sealing has a flat disk or closure member that provides sealing by means of the fluid pressure. Slab gate valves, unlike expanding and wedge gate valves, are not torque seated.

Gasket: A gasket is a mechanical seal that fills the space between two mating surfaces to prevent leakage.

Gland: A valve component that is installed on the packing to transfer the compression load from the gland flange to the packing. A gland is a removable part of the valve that should be removed from the stem area during packing repair and maintenance. A gland may be combined with a gland flange, as illustrated in Fig. 1.13. The gland is fastened and attached to the valve bonnet with gland bolts and nuts as illustrated in Fig. 1.14.

Gland bolt torque: The amount of torque (defined as the required force multiplied by the distance in newton meter (N.m) or lbf-ft or lbf-inches (lbf is 1-pound force)) required to fasten the bolt and nut to the gland flange. The torque or load amount applied to the gland flange is important, since it exerts axial load on the packing rings to seal the stem area.

FIGURE 1.13 Combined gland and gland flange on packing rings. *Courtesy: Magazine/Elsevier.*

Gland flange: A flange-shaped component (see Fig. 1.14) located on the top of a gland or gland follower to retain and enforce the gland into the packing or stem sealing. Gland flange contains two holes for bolting. The bolts should provide a sufficient compression load on the gland flange, which is eventually transferred to the packing through a gland. The compression load on the packing is transferred into radial force from the packing to the bonnet and stem to provide sealing. Fig. 1.14 illustrates a gland flange with two bolts on a small wedge gate valve.

FIGURE 1.14 Gland flange and bolting on a wedge gate valve. *Courtesy: Shutterstock.*

Global warming: Global warming is the gradual heating of the earth's surface, oceans, and atmosphere caused by air pollutants such as methane and carbon dioxide and the greenhouse effect.

Globe valve: A type of valve used for flow regulation or throttling. As illustrated in Fig. 1.15, the fluid makes two 90° turns inside the valve, which creates significant pressure drop inside the valve. This significant pressure drop may lead to wearing and erosion of the valve internals. Cavitation is one of the main operational problems affecting globe valves in liquid services due to this significant pressure drop. Globe valves are subject to fugitive emission testing based on API 624 and ISO 15848-1.

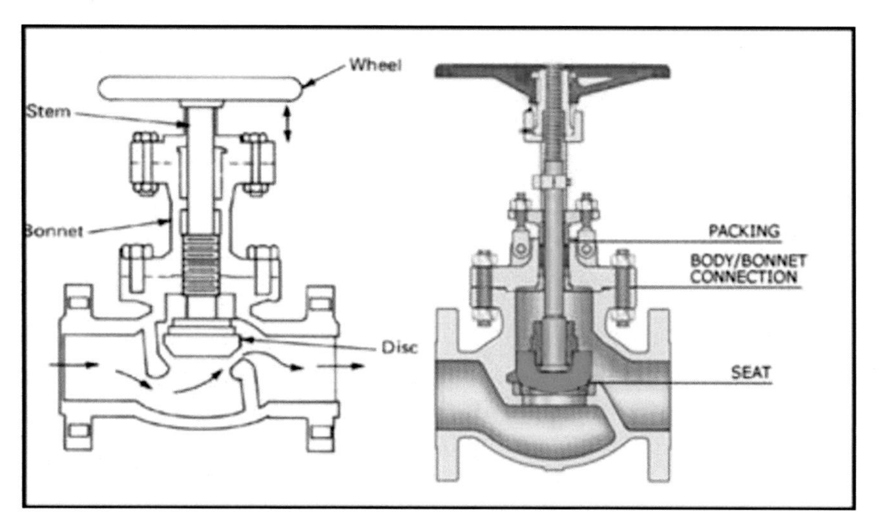

FIGURE 1.15 Globe valve.

Greenhouse effect: The greenhouse effect is a natural process that warms the earth's surface. When the Sun's energy reaches the earth's atmosphere, some of it is reflected back to space and the rest is absorbed and reradiated by greenhouse gases.

H

Hastelloy: Hastelloy is a corrosion-resistant nickel alloy that contains other chemical elements such as chromium and molybdenum. This material has high temperature resistance and exceptional corrosion resistance. Hastelloy C276 is one of the grades of Hastelloy used in the oil and gas industry.

Hazardous air pollutants (HAPs): HAPs, also referred to as toxic air pollutants, are those that can cause serious health problems such as cancer. The EPA has detected 187 types of hazardous air pollutants, such as lead, ozone, nitrogen dioxide, etc.

High strength bolting: As per ASME B31.3 "Process piping code," bolts with a yield strength of more than 30 Ksi are considered high yield or high strength bolts. High yield bolts should be used for high-pressure class piping. Piping systems with a pressure value of minimum CL600 (CL stands for class) equal to a pressure nominal (PN) of 100 bar are considered high-pressure class piping.

Hydrogen sulfide: Hydrogen sulfide (H_2S) is a chemical compound with two atoms of hydrogen and one atom of sulfur (see Fig. 1.16). It is an undesirable by-product of oil that is corrosive, extremely toxic, and flammable. A concentration of this compound as low as 50 ppm can kill a human being. Hydrogen sulfide has a smell like a rotten egg.

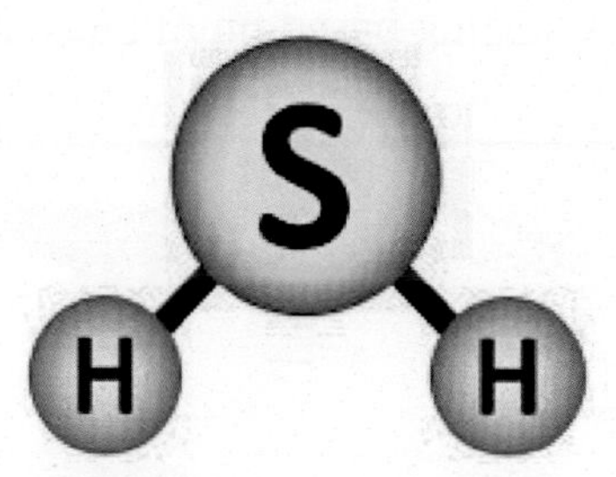

FIGURE 1.16 Hydrogen sulfide.

I

ISO: The International Organization for Standardization is a standard-setting body that establishes many standards, especially in the oil and gas industry. ISO 15848-1 is the main standard addressing fugitive emission testing and qualification for industrial valves.

Isolation valve: An isolation valve, also called an isolating valve, is a type of valve used for fully closing and fully opening the flow path. Gate, ball, and butterfly valves are categorized as the main isolation valves used in oil and gas industry.

J

Joint efficiency: Joint efficiency can be defined as the reliability obtained from joints after welding. Joint efficiency can be calculated as the ratio of the weld strength to the strength of the base material. Higher joint efficiency can reduce the risk of fugitive emissions from welded connections such as piping.

K

Kyoto Protocol: Put simply, the Kyoto Protocol is a United Nations framework for climate change that defines the commitment of industrial countries to reduce and limit greenhouse gases and fugitive emissions to the agreed-upon amounts.

L

Leachable: Compounds such as chloride and fluoride in a graphite packing are known leachable. Leachable in the valve stem packing should be limited to prevent corrosion and leakage from the packing.

Leak/Leakage: The amount of fluid escaping from the packing and gland or through body seals during the fugitive emission test or operation.

Leakage rate: The amount or quantity of fluid passing through the sealing or packing during a specific period.

Leak detection and repair (LDAR): An LDAR program is a systematic procedure that is utilized to identify and repair leaking components, including valves, pumps, connectors, and compressors, to minimize the fugitive emission of vapor organic compound (VOC) and HAPs. Successful implementation of LDAR involves five steps that are explained in detail in Chapter 2.

Lip seal: A type of sealing used as a stem seal or seat and body seal for industrial valves. The seal is made of two parts; one part is a soft material such as PTFE (Teflon) and the second part is a metallic spring in a material such as Inconel 625 or Elgiloy (Cobalt alloy) that energizes the soft material part. Fig. 1.17 illustrates lip seals made of PTFE in white and with Inconel 625 springs in gray color.

FIGURE 1.17 Lip seals.

Liquid penetrant test: A liquid penetration test, also called a penetrant test (PT) or dye penetration inspection, is a very common nondestructive test (NDT) in the valve industry based on using three types of liquids. The first liquid is a penetrant, which is a red liquid or spray; the second is a developer, which is a white liquid; and the last is a cleaner.

Liquefied natural gas (LNG): LNG is a type of natural gas that can be cooled down to a very low temperature such as −260°C to be stored and sold as a type of liquid at higher prices compared to natural gas. LNG typically contains methane with some mixture of ethane C_2H_6.

Live load packing: Also called "live load gland flange." The simplest form of live loading is the application of a spring that is placed on the gland flange, gland, and packing. The spring provides additional loads on the packing for tighter stem sealing. A live load spring provides sufficient load on the gland flange and packing even if the gland flange load is not sufficient after some time of operation. In fact, live load spring compensates for lack of gland flange load on the packing due to a variety of reasons such as misalignment of the gland flange. Fig. 1.18 illustrates a small gate valve with two springs on the gland flange to provide live load on the gland flange and packing.

FIGURE 1.18 Live load spring on a small gate valve gland flange. *Courtesy: Elsevier.*

Loctite: A German brand of sealant used in the valve industry on vent and drain plugs to prevent leakage from the plugs during operation. Fig. 1.19 illustrates the use of Loctite on plug threads that will be installed on the body of a ball valve to prevent leakage of fluid from the plug threads during operation.

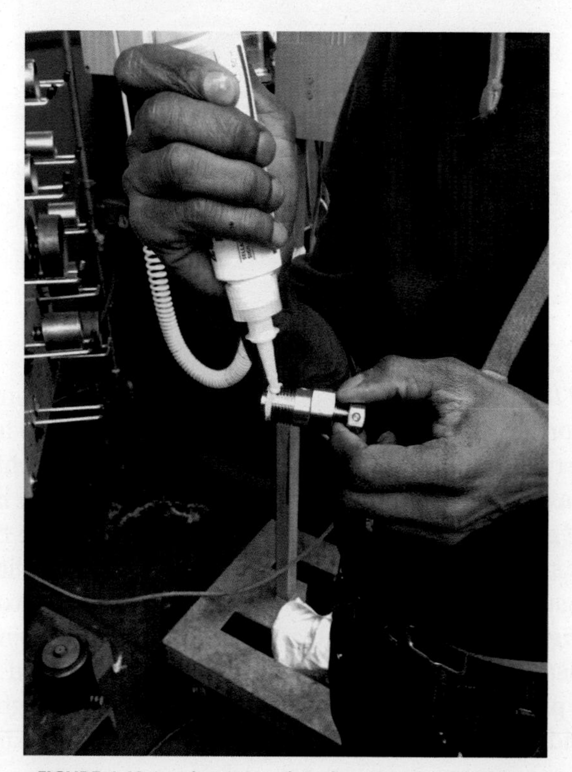

FIGURE 1.19 Loctite on a valve plug. *Courtesy: Elsevier.*

Low emission valve: Any kind of valve that is designed and produced in such a way as to provide no or low fugitive emissions as per the limits provided in the international standards such as API and ISO. A low emission valve should pass the required low fugitive emission tests.

M

Magnetic particle test (MPT): MPT is a type of NDT process for detecting the defects on the surface by using magnetic particles.

Martensitic stainless steel: Martensitic stainless steel is a kind of stainless steel characterized by hardness and high mechanical strength. The heat treatment of this type of stainless steel can involve age hardening and tempering, which leads to greater hardness of the material. 13 chromium, also known as grade 410, is one of the most important martensitic stainless steels used for valve stems in the oil and gas industry.

Mechanical cycle: A motion of the valve stem simulating the movement of the valve closure member or obturator such as the ball or disk from fully closed position to fully open position and returning to fully closed position.

Methane: Methane (CH_4) is the lightest and simplest form of hydrocarbon. Methane is a powerful greenhouse gas. This gas is flammable and is used for fuel worldwide. It is a colorless gas and does not have any smell at low concentration. However, it has a sweet smell at high concentration. The mixture of methane in air with a concentration above 5% could be explosive. Methane is considered a kind of volatile organic compound.

N

Nickel alloy: Compound steels with nickel as the principal element, such as Inconel 625 and Hastelloy C276, are considered nickel alloys. Nickel alloys in general are expensive, corrosion-resistant alloys that are used for different valve components.

Nominal pipe size (NPS): The size of piping or valves in inch as an imperial unit. NPS 2 means that the size of the piping is 2″.

Non-destructive testing (NDT): NDT refers to a group of tests for materials and components performed to detect possible defects without causing damage to the material. NDT is typically performed on weld joints, and could include different types of tests, such as visual inspection, magnetic particle tests, liquid penetration tests, radiography, and ultrasonic tests. Ultrasonic test and radiography NDT are categorized as volumetric NDT.

NORSOK: NORSOK refers to a set of standards developed by the Norwegian petroleum industry to ensure adequate safety, value adding, and cost-effectiveness for petroleum industry developments and operations.

O

O-ring: A type of valve seal used to block the fluid inside the valve and prevent fugitive emission. O-rings are typically placed inside a groove between two metallic surfaces. O-rings can be used for valve stem seals or between the valve body and seat.

Obturator: Also called a valve closure member, an obturator is a part of a valve positioned inside the flow path to permit or prevent the flow. Different valves have diverse shapes and types of obturators, such as a ball, gate, disk, or wedge.

Oil reservoir: Also called a "petroleum reservoir" or "oil and gas reservoir," an oil reservoir is a subsurface area or pool in which hydrocarbons (oil and gas) accumulate in a porous rock formation.

Operator: A device or assembly used to open and close a valve. The valve could be manually operated via a handwheel and gear box or lever. Alternatively, the valve could be automatically operated by an actuator plus a control panel. A control panel controls the fluid input such as hydraulic or pneumatic power to the actuator. Hydraulic and pneumatic actuators are typically operated with a control panel, while electrical actuators are operated by a motor and do not require any control panel.

P

Packing: Packing is a type of a valve sealing in soft (nonmetallic) materials such as graphite or Teflon (PTFE) located between the valve stem and bonnet to prevent leakage from the stem area of the valve. Packing provides sealing by the force applied to it from the top by the gland follower. Fig. 1.20 illustrates a stem packing arrangement.

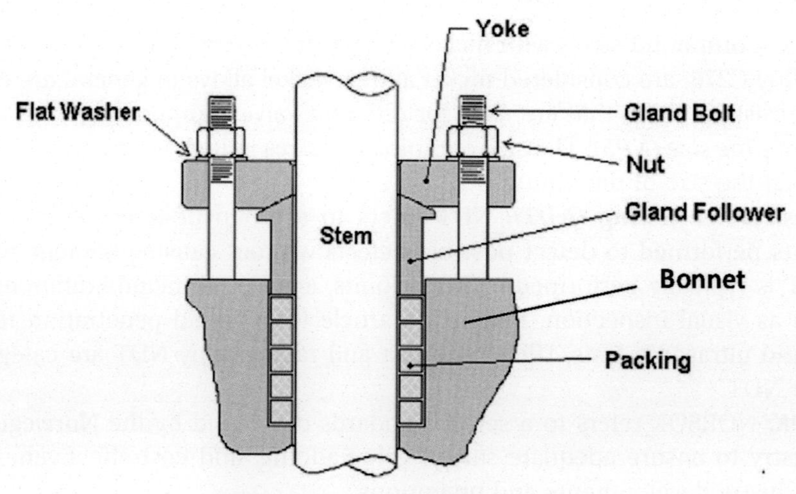

FIGURE 1.20 Stem packing arrangement.

Packing size or cross section: Two sizes of packing, $^1/_4$ ″ and 1/8″, are provided in API 622. Fig. 1.21 illustrates a valve stem, valve stuffing box (shown in blue), and the area between the stuffing box and the stem where the packing sits. In fact, the packing fills the area between the stem and bonnet. Packing size or cross section is calculated through Formula 1.1.

$$\text{Packing size} = \frac{\text{Stuffing box internal diameter} - \text{Stem outside diameter}}{2} = \frac{B - A}{2} \qquad (1.1)$$

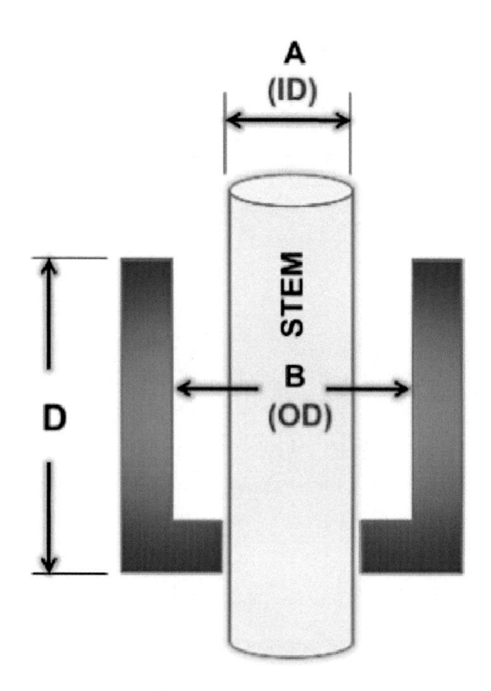

FIGURE 1.21 Stem, stuffing box, and packing arrangement.

Question: The stem diameter for a gate valve undergoing a fugitive emission test is 11.1 mm. The stuffing box diameter for the valve is 17.52 mm. What packing size should be used for this valve?
Answer: Stem diameter = 11.1 mm = 0.437 in.; Stuffing box diameter = 17.52 = 0.690 in.

$$\text{Packing size} = \frac{0.690 - 0.437}{2} = 0.1265 \sim 0.125 \text{ in.} = 1/8'' \text{ packing size}$$

PEEK: PEEK stands for polyether ether ketone, an organic, thermoplastic material normally used as a soft seat or soft seal between the back seat of a valve and its body. PEEK is compatible with chemicals and process fluid and is suitable for high-pressure classes such as ASME pressure class 2500 or even higher. PEEK can be used in a wide range of operating temperatures from −80°C to 200°C.

Pipe loads: Different types of loads are applied on the piping system, some of which can cause leakage. Fig. 1.22a illustrates axial loading on the piping and connected valve. Axial loading is a force directed over the piping on an axis. Fig. 1.22b illustrates bending moment stress on the piping and connected valve, which causes the piping to bend. Fig. 1.22c illustrates the torsion stress that is produced when a twisting force or moment is applied to the end of piping.

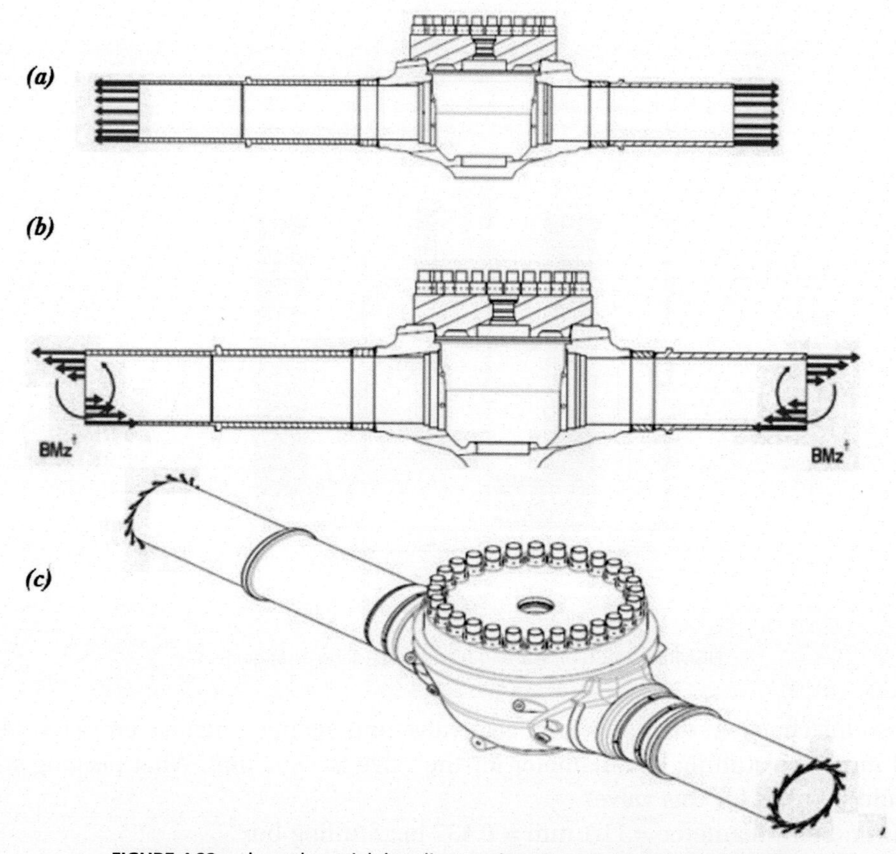

FIGURE 1.22 a, b, and c: axial, bending, and torsion loads. *Courtesy: Elsevier.*

Pipe stress analysis: Pipe stress analysis predicts stresses in the pipe as well as loads on the connected equipment due to different factors such as weight, pressure, temperature changes or shocks, etc. Various software is available on the market for performing stress analysis, such as CAESAR II, Triflex, etc.

Piping and instrument diagram (P&ID): A diagram that shows the process of fluid including the interconnection of process equipment and the instrumentation used to control the process. In fact, P&IDs are more detailed compared to process flow diagrams.

Pitting corrosion: Pitting corrosion is a form of localized corrosion by which cavities or holes are produced inside the material. This type of corrosion typically occurs in chloride-containing environments such as seawater. Fig. 1.23 illustrtaes pitting corrosion on a face of flange used in oil and gas industry.

FIGURE 1.23 Pitting corrosion in a flange. *Courtesy: Shutterstock*

Poisson's ratio: An expression in material mechanics in which the Poisson effect is measured. The Poisson effect is defined as a material's tendency to expand in a direction perpendicular to the compression direction. Poisson's ratio is calculated through Formula 1.2.

$$\mu = -\frac{\varepsilon_{trans}}{\varepsilon_{axial}} \tag{1.2}$$

where:

μ: Poisson's ratio;
ε_{trans}: Transverse strain;
ε_{Axial}: Axial strain.

Polymer: A polymer, as illustrated in Fig. 1.24, is a chemical compound with molecules bonded together in long, repeating chains. The term polymer is widely used in the plastic industry. One of the main characteristics of a polymer is elasticity. Thus, elastomers are a type of polymers (c.f. elastomer). Synthetic and natural rubbers are considered a type of elastomer.

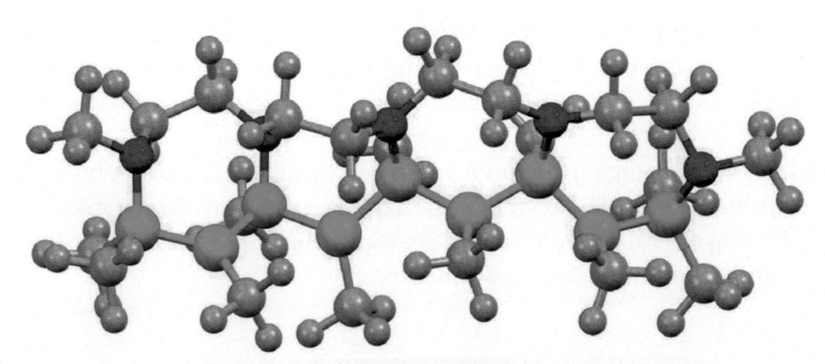

FIGURE 1.24 Polymers.

Polytetrafluoroethylene (PTFE): PTFE is a synthetic fluoropolymer of tetrafluoroethylene. This material is used for valve seats, mainly those of ball valves, and in stem packing. PTFE has good temperature resistance and a smooth surface.

ppm: Ppm stands for parts per million; 1 ppm means one part per 10^6 parts or 1,000,000 parts. As an example, 100 ppm means 100 parts per 1,000,000 parts, which is 0.1%. Similarly, ppmv means the measurement of parts per million volume.

PREN: PREN stands for pitting resistance equivalent number, which is calculated according to Formula 1.3 to assess the resistance of stainless steel against pitting corrosion according to chemical composition percentage. Typically, a PREN value above 40 indicates good resistance against seawater corrosion.

$$\text{PREN} = \%\ \text{Chromium} + 3.3\%\ \text{Molybdenum} + 16\%\ \text{Nitrogen} \tag{1.3}$$

Pressure class: A rounded number such as 150, 300, 600, 900, 1500, 2500, or 4500 as per ASME standards such as ASME B16.5 used for flange design, and ASME B16.34 used for valve design. Each pressure class represents a pressure value or pressure nominal (PN) at a specific temperature. As an example, class 150 represents PN 20 bar at ambient temperature. In addition to ASME pressure classes, API has pressure classes, such as 5000 psi, 10,000 psi, and 15,000 psi, that are defined in standards such as API 6A, "Specification for wellhead and Christmas tree." Typically, valves installed on the wellhead and those located underwater, called subsea valves, follow API pressure classes, which are in general higher than ASME pressure classes. ASME pressure classes are common for topside valves and for refineries and chemical plants. The ASME pressure class is typically noted on the valve or pipe by writing the word "class" or "CL," followed by the pressure class number (e.g., class 300 or CL300).

Pressure-containing components: Sections or parts of a valve whose failure to function leads to the leakage of the internal fluid to the environment. Body, bonnet, stem, and bolting are categorized as the pressure-containing components of valves.

Pressure-controlling components: Sections of the valve used to control or regulate the movement of the pressurized fluid inside the valve, such as valve internals like the closure member and seats.

Pressure nominal (PN): A term used to describe the pressure value in bar that a pipe is designed to withstand. More precisely, a pipe's PN number describes the nominal pressure that a pipe can support at ambient temperature. As an example, PN 20 indicates that the pipe and connected valve can withstand 20 bar.

Pressure relief valves: A relief valve or pressure relief valve is a type of safety valve used to control or limit the pressure in a system; pressure might otherwise build up and create a process upset, instrument or equipment failure, or fire. Fig. 1.25 illustrates a typical pressure relief valve. The overpressure fluid enters the valve from the inlet port, pushes the disk upward by overcoming the spring force, and exits the valve through the outlet port. Unlike a gate, globe, or ball valve, a pressure relief valve does not have any stem, stem seal, or stem packing.

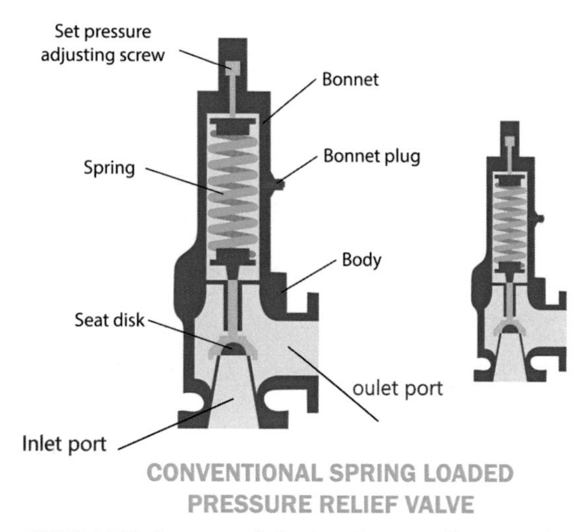

FIGURE 1.25 Pressure relief valve. *Courtesy: Shutterstock.*

Process flow diagram (PFD): A PFD is a diagram commonly used in the oil and gas industry and generated by the process department to show the main process, including the main piping and equipment. A PFD shows the process of oil, gas, and water treatment in general in oil field developments. In addition, PFDs in refineries and petrochemical plants illustrate how the feed fluid service that enters the plant is converted to the final product. Fig. 1.26 illustrates a PFD showing the gas input to a compressor station and a knockout drum installed after the compressor to separate the liquids from the compressed gas generated in the compressor.

Psig: The abbreviation of pound per square inch gauge or the amount of force in pounds that is applied on a square inch. Psig refers to the pressure measured by a

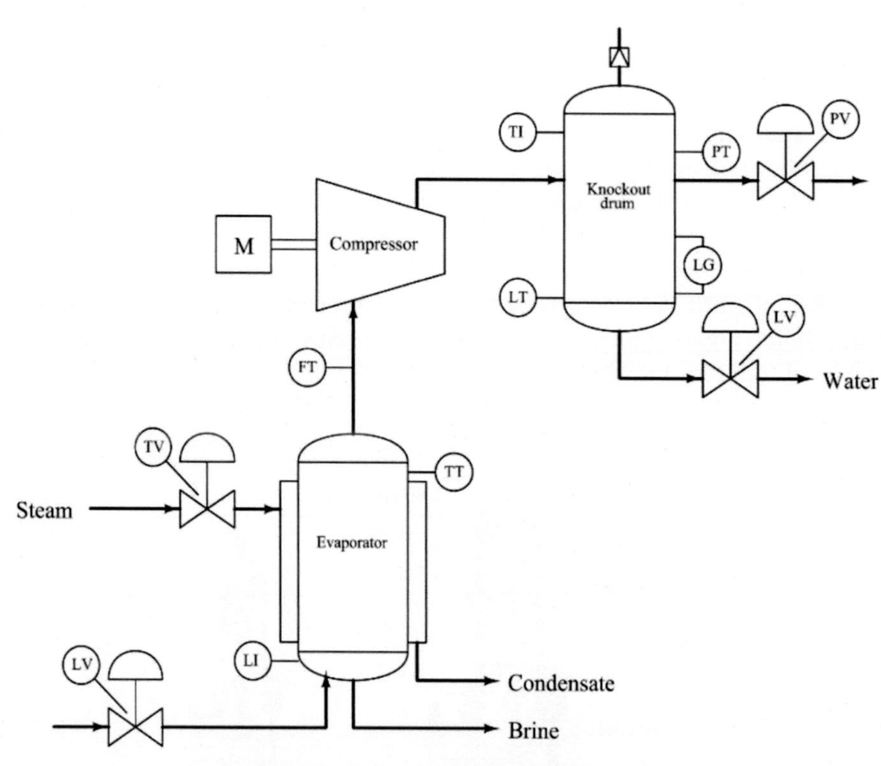

FIGURE 1.26 Process flow diagram (PFD) sample.

pressure gauge. The pressure gauge in Fig. 1.27 illustrates a pressure value between 800 and 900 psig. The pressure in the system, as illustrated in Fig. 1.28, is the total of the atmospheric pressure, which is 1 bar equal to 14.7 psi, plus the gauge pressure as per Formula 1.4. It should be noted that 1 ksi or kilo-psi is equal to 1000 psi.

Relationship between gauge pressure and absolute pressure

$$\begin{aligned} &\text{Absolute Pressure} = \text{Gauge pressure} + \text{Atmospheric pressure} \\ &\text{Atmospheric pressure} = 1\ \text{bar} = 14.5\ \text{psi} \end{aligned} \tag{1.4}$$

FIGURE 1.27 Pressure gauge.

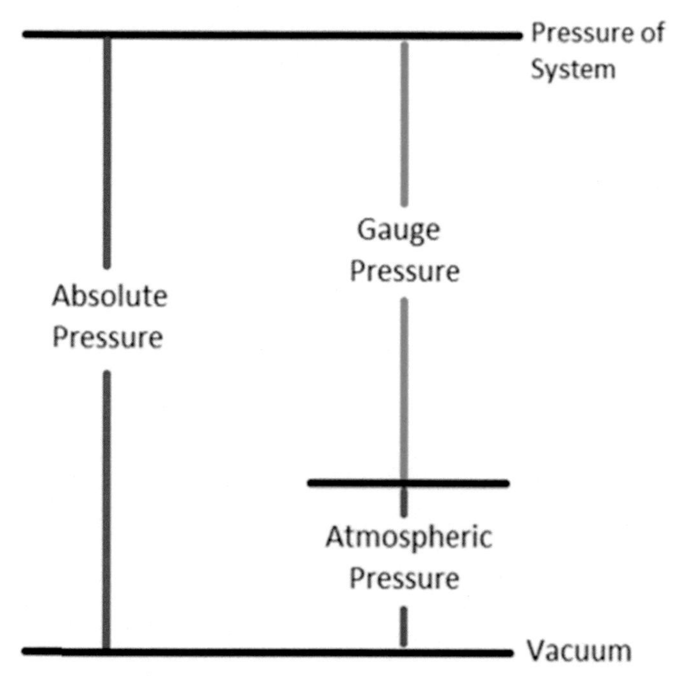

FIGURE 1.28 Relationship between pressure of the system in psi and psig.

Pump: A piece of mechanical equipment or facility used in the oil and gas industry to pressurize and move liquids such as oil and water forward. Pumps are categorized in different types such as rotary and reciprocating. A discussion of the various types of pumps is beyond the scope of this book.

Q

Quarter turn: Valves whose stem and closure member rotates just 90° between opening and closing positions. Ball and butterfly valves are categorized as quarter-turn valves.

R

Radiography test: A method of nondestructive test in which X-rays or gamma rays are used to detect possible defects or flaws in components such as weld joints.

Rapid gas decompression (RGD): See Anti-explosive decompression. Explosive decompression (also referred to as rapid gas decompression or RGD) is a failure mechanism of elastomer seals and O-rings that is caused by a rapid reduction in the pressure of a gaseous media. Gas that has permeated into the elastomer seal expands violently when the pressure is released rapidly, causing fissuring and seal failure.

Ra roughness: As described in ASME B46.1, the average of a set of individuals' measures of roughness peaks and valleys on a specific length (L) as illustrated in Fig. 1.29. Rq roughness is the root mean square average of the profile height over the evaluation length.

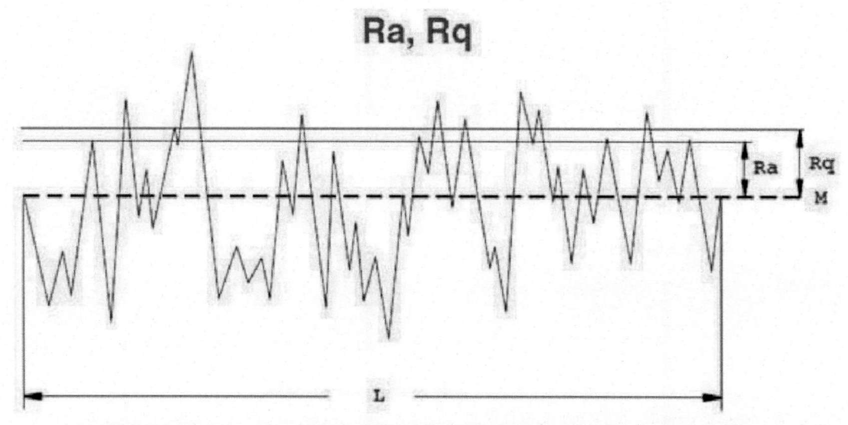

FIGURE 1.29 Ra roughness as an average value between peaks and valleys.

Reynold number: The ratio of inertial forces to viscous forces and an important factor to designate the flow condition as laminar or turbulent (see Fig. 1.30). Laminar flow is when the Reynold number is less than 2000 and the fluid velocity is low. Turbulent flow is associated with a Reynold number above 4000 and the fluid velocity is high. Formula 1.5 provides a formula for calculating the Reynold number.

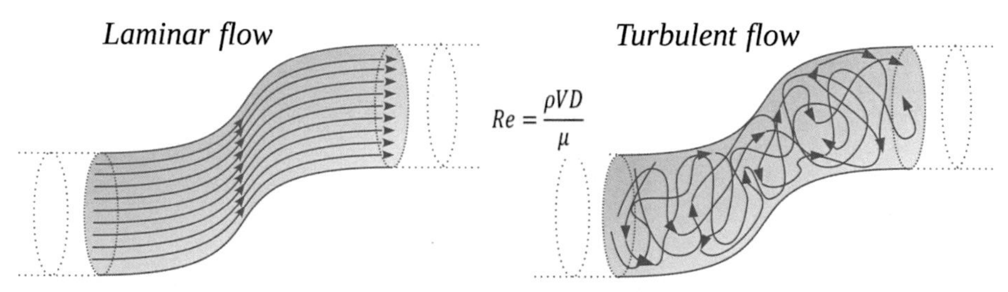

FIGURE 1.30 Reynold number formula, and laminar and turbulent flow. *Courtesy: Shutterstock.*

Where:

ρ: Fluid density in kg/m^3;
V: Fluid velocity m/s;
D: Pipe diameter in m;
μ: Fluid dynamic viscosity kg/m.s.

Rising stem valve: A valve in which the movement of the stem is linear and axial up- and downward without any stem rotation. Gate and globe valves could be categorized as rising stem valves.

Rotating stem valve: A valve, such as a ball valve, that is opened and closed through a 90° rotation of the stem.

Rising–rotating stem valve: A valve in which the movement of the stem is the combination of rising, linear up- and downward movement and 90° rotation.

S

Schedule: Schedule, also called pipe schedule, is a term used to describe the thickness of a pipe. Different schedule numbers, such as 40, 60, 80, etc., are defined in piping standards such as ASME B36.10, which covers the standard dimensions for carbon steel piping. Higher schedule number indicates thicker piping.

Seals: Valve seals are materials that are placed between two main valve components, such as the body and bonnet, stem and body, stem and bonnet, or seat and body, to prevent leakage from the valve. Seals can be metallic, nonmetallic, or a combination of both. Fig. 1.31 illustrates triple O-rings in black color around a ball valve stem.

Seamless pipe: Seamless refers to a method of piping manufacturing in which no welding is applied. Therefore, the joint efficiency of seamless piping in general is higher than welded piping. The manufacturing process involves stretching and making a seamless pipe from a billet. The billet that is formed initially is made of raw steel. Steel

FIGURE 1.31 Triple O-ring around the stem of a ball valve.

billets, illustrated in Figs. 1.32, are like bars or pipes without any hole inside. The billets are heated up and a mandrel is passed through them via a rolling process to make seamless piping as illustrated in Fig. 1.33.

Seat insert: A seat insert, also called a seat, is a soft ring in PTFE or PEEK that is placed in the seat carrier or retainer. The seat insert provides sealing between the seat retainer and closure member in a ball valve.

Shaft: See Stem.

FIGURE 1.32 Billets.

FIGURE 1.33 Piercing a heated-up billet with a mandrel.

Side-entry valve: Unlike top-entry valves, the assembling and disassembling of a side-entry valve is performed from the side. Online repair and maintenance of this type of valve is not possible; it must be removed from the connected piping for maintenance. Fig. 1.34 illustrates a side-entry ball valve including three body pieces.

Slip-stick: Slip-stick, also called stick-slip, is a phenomenon in which a spontaneous jerking motion occurs when two objects slide over each other. Slip-stick can occur between the stem and contact packing if the surface finish of the stem in contact with the packing is too low.

FIGURE 1.34 Side-entry ball valve. *Courtesy: Perar.*

Socket weld: This type of welding is a type of piping and valve connection in which the pipe is inserted into the recessed or female area of the valve or other piping components to be welded. A socket weld is typically used for piping in sizes of 2″ and smaller. This type of connection is not common in the offshore sector of the oil and gas industry. Fig. 1.35 illustrates a socket weld between a pipe (Item #3) and a socket weld flange (Item #1). Item #2 is a fillet weld between a pipe and flange. One of the problems associated with socket welding is crevice corrosion, which can occur in the areas shown in green in Fig. 1.35 and highlighted with the letter X.

SSA: SSA stands for stem sealing adjustment during the fugitive emission tests. In some cases, no stem sealing adjustment is allowed in API 622 and API 624. One stem sealing adjustment is allowed in API 641 during the test. ISO 15848-1 allows one, two, or three stem sealing adjustments depending on the number of cycles.

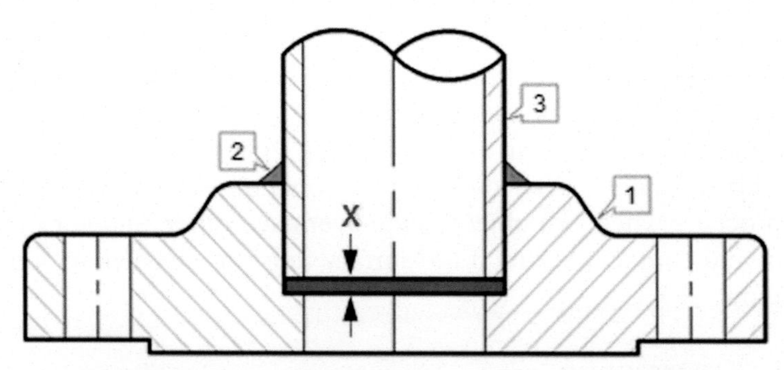

FIGURE 1.35 Socket weld connection between a pipe and flange.

Stainless steel: A group of iron-based alloys that contain a minimum of approximately 11% chromium. Stainless steel grade 316 is a type of austenitic stainless steel that has approximately 18% nickel, 8% chromium, and 2% molybdenum. Austenitic is a category of stainless steel that has less mechanical strength compared to other types.

Stem: The stem or shaft is the part of a valve that transfers the load from the valve operator device, which could be a gear box or an automatic component (actuator), to the valve internals.

Stem nut: Also called bushing or a yoke nut. A stem nut is typically a one-piece component installed on the top section of a valve stem that transfers the torque or rotational movement to linear movement of the stem, mainly in gate and globe valves. Fig. 1.36 illustrates the position of a stem nut highlighted in yellow on the top part of the stem of a gate valve. Locking the stem nut around the stem leads to the generation of linear movement for the stem in upward and downward positions even though the operator provides rotary forces for the stem movement.

Stem seal: A stem seal consists of components such as O-rings, lip seals, or graphite packings that are installed around the stem to prevent fugitive emission and leakage from the stem seal area to the environment. The other name for stem seal is "shaft seal."

Strain: Deformation of a material due to stress, calculated from the ratio of change in the length to the original length of the material.

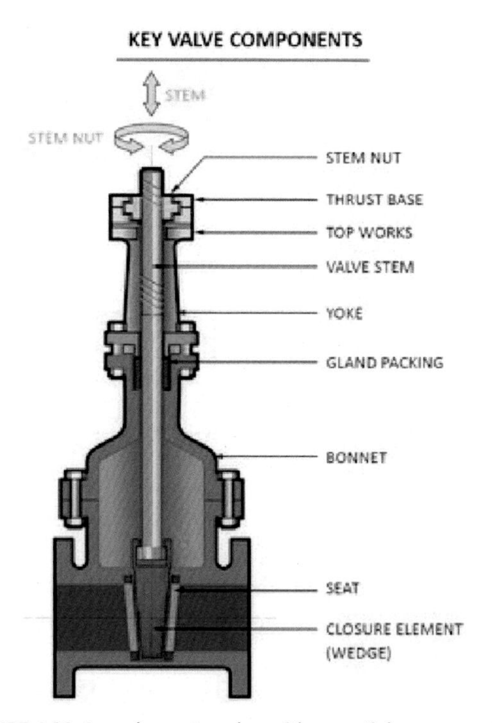

FIGURE 1.36 A wedge gate valve with essential components.

Stress corrosion: Corrosion types that are initiated and exacerbated in combination with stress. One type of stress corrosion is chloride stress cracking corrosion, which can occur in marine environments. The other type is sulfide stress cracking corrosion which occurs due to hydrogen sulfide.

Stuffing box: Also called a packing chamber, a stuffing box is a space between the valve stem and bonnet in which the compression packing is inserted. Fig. 1.37 illustrates a stuffing box area packing arrangement.

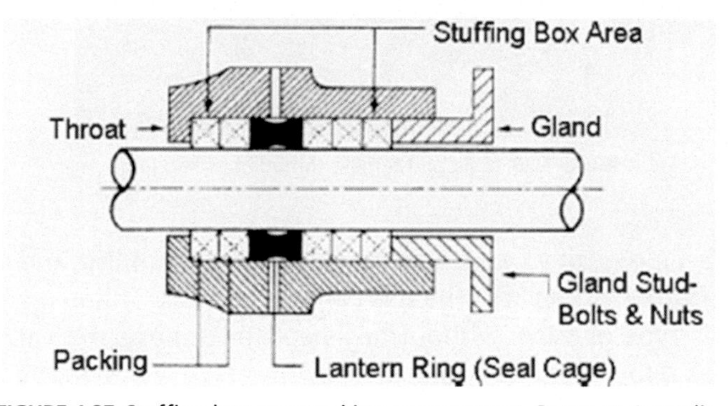

FIGURE 1.37 Stuffing box area packing arrangement. *Courtesy: Australia.*

Stuffing box diameter: Also called stuffing box bore, stuffing box diameter is defined as the inside diameter into which packing is inserted.

Surface finish: A measurement of the roughness of a surface, which is typically measured and expressed in micrometers or microinches. A lower amount of surface finish means a smoother surface with less roughness.

T

TA Luft: *Technische Anleitung zur Reinhaltung der Luft* in German, translated as "Technical Guidelines for Air Pollution Control" in Germany.

Tensile load: The ratio of the applied force divided by the unit area that has a tendency to elongate or stretch the material. Tensile load, which tends to increase the length of a material, is the opposite of compression, which tends to reduce the length of a material. Tensile load is also called tensile stress.

Thermal cycle: A change in temperature from ambient temperature to an elevated temperature and then back to ambient temperature.

Thermoplastic: Thermoplastic materials are a group of plastic polymers that become moldable at a certain elevated temperature. PTFE is a type of thermoplastic material used for valve seats and stem packing.

Threaded connection: Threaded joints are detachable joints that are connected directly through male and female threads. Threaded joints can be used for small piping, typically in sizes of 2″ or less. Threaded connections are subject to corrosion and fatigue due to loads and are not common in the offshore sector of the oil and gas industry. Fig. 1.38 illustrates male and female threaded piping connections.

FIGURE 1.38 Male and female threaded piping joints.

Top-entry valve: A type of valve design in which the assembling and disassembling of the valve is done from the top part, which is called a bonnet. Online maintenance can be performed on this type of valve without any need to remove the valve from the line. Fig. 1.39 illustrates a top-entry ball valve design.

FIGURE 1.39 Top-entry ball valve design.

Torque: Torque is the twisting force that causes rotation. Valve manufacturers provide a torque table for the valves indicating the amount of torque required for valve operation during the opening and closing of industrial valves. Torque can also be defined as the amount of twisting force multiple rotation distance provided by the bolting (bolts and nuts) on the gland and gland flange as well as the packing.

Trim: Internal parts of a valve that are in contact with fluid, such as the closure member and seats, are called trim. Trim should have as a minimum the same material as the body, and the material of the trim should be in a corrosion-resistant alloy.

Type testing: A test that is conducted to assess the performance of a specific valve design at a specific temperature and pressure.

U

UNS: UNS stands for unified numbering system. It is an alloy designation system widely accepted globally, and especially in North America. Each material has a unique UNS that is connected to its chemical composition and mechanical properties.

Upstream oil and gas: The oil and gas industry can be divided into two sectors: upstream and downstream. The upstream sector refers to exploration, well drilling, and production. Subsea and offshore oil and gas production can be categorized as belonging to the upstream sector, while refineries and petrochemical plants belong to the downstream sector of the oil and gas industry.

V

Valve: A valve or industrial valve is a mechanical component used in a piping system to open and close the fluid passage, prevent backflow of the fluid, and control the flow. Valves may also be used for safety purposes.

VDI 2440: Association of German engineers, guideline 2440.

Vee pack or V-pack: This type of sealing is a very high-tech, advanced sealing that can be used for stem sealing as well as sealing between the valve closure member and body. Vee pack sealing (illustrated in Fig. 1.40), also called Chevron V-pack sealing, includes male and female adaptors or layers that could be a group of three or five layers or even more. This type of sealing can withstand very high pressure and heavy-duty applications like the valve stem sealing or body and closure member sealing of subsea valves. Fig. 1.40 shows a V-pack made of PEEK and PTFE thermoplastic seals. A V-pack could be made from layers of elastomers such as Viton FKM, Neoprene, EPDM, etc. The bottom and top layers have one flat side surface on both ends, rather than the V shape, with other V-shaped layers located in between. The combination of construction materials defines the resistance of the V-pack against pressure, temperature, and fluid. V-packs are also suitable sealing for rods and pistons.

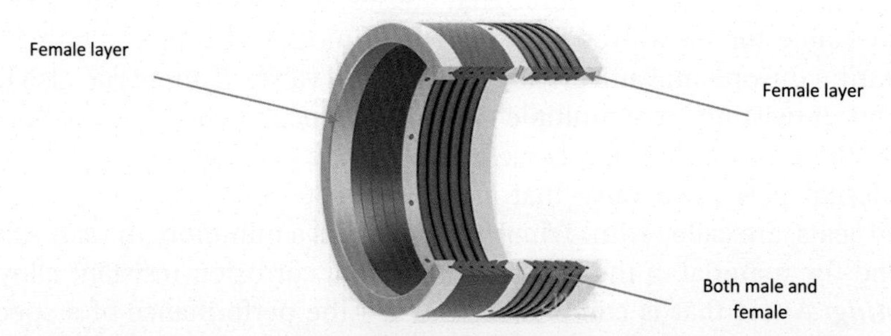

FIGURE 1.40 V-pack including male and female adopters made of PTFE and PEEK.

Vent: Vent or gas venting is the intentional and controlled release of useless and unwanted gasses such as methane to the atmosphere. The gas vent process is similar to the drain process for liquids.

Viton: A type of synthetic rubber and fluoropolymer elastomer used for valve seals in the shape of an O-ring. Viton has very good chemical resistance and can be used for high pressure and temperature applications.

Visual testing or inspection: This is the simplest and probably the oldest approach to inspecting a component or material by looking at it directly or using instruments such as a magnifying glass, borescope, or mirror.

Volatile gas: A gas substance that changes its status from gas to the liquid at very low temperature and high pressure. Methane is a gas that is extremely volatile at room temperature. Typically, methane can be liquefied at very low temperature conditions such as −150°C to −161°C, roughly speaking.

Volatile organic compound (VOC): VOCs are organic compounds or chemicals that have high vapor pressure at ordinary room temperature and low water stability. Many VOCs are synthetic chemicals used in the industry. VOCs are typically emitted as gases from solids or liquids.

W

Wear: The interaction of the surface of a material with the environment or another material, which leads to the removal of some metal from the material surface.

Y

Yield: Yield, also called yield stress or yield strength, is the amount of force required to be applied on an object to cause it to change from elastic to plastic. Yield stress is defined for steels and components that are made in steel as the maximum stress that is applied to the steel before permeant shape change occurs.

2 Fugitive emissions from piping and valves

2.1 Introduction to fugitive emission

Fugitive emission is defined as the unintentional and undesirable emission, leakage, or discharge of gases or vapors from pressure-containing equipment or facilities, and from components inside an industrial plant such as valves, piping flanges, pumps, storage tanks, compressors, etc. Fugitive emission is also known as leak or leakage. The term "fugitive" is used because these emissions are not taken into account and calculated during the design of the equipment and components. In addition, these emissions are unanticipated; as such, they are not detected by typical monitoring and control devices. Since the typical control equipment and facilities cannot detect them, fugitive emissions can also be called uncontrolled or unanticipated emissions.

Humans have been polluting the environment for hundreds of years, especially since the advent of the industrial revolution in the 19th century. The major increase in fugitive emissions started approximately 70 years ago in the 1950s, when the world began to require significantly more heat and electricity. It is noteworthy that the amount of fugitive emission in 2020 is almost five times more than it was 70 years ago. Due to growing concerns about health, safety, and environmental protection, fugitive emission has become a key concern for end users, operators of oil and gas, chemical and petrochemical plants, as well as regulators across the globe. A variety of reasons and factors such as regulations from governments, health, safety, and environment (HSE) programs, and increasing pressure from the public have forced end users in the oil and gas industry to become more conscious of and sensitive to decreasing fugitive emission. Different regulations and standards have been developed, starting in the early 1960s, along with regulatory bodies and legislation such as TA Luft in Germany and the Clean Air Act in the United States. Regulations for emission prevention and measurement are constantly under development for the reasons mentioned above, as well as pressure from society. The standards for valve fugitive emissions are included in Chapter 4.

Unfortunately, it is difficult to quantify the exact amount of fugitive emission produced by different countries with a high degree of accuracy. As an example, there is a high degree of uncertainty about the level of fugitive emission in some of the major oil- and gas-producing countries, such as Russia, a member of OPEC (Organization of the Petroleum Exporting Countries). The evaluation of fugitive emission is a climate issue, but it is also a political and economic concern.

Prevention of Valve Fugitive Emissions in the Oil and Gas Industry. https://doi.org/10.1016/B978-0-323-91862-6.00002-2

2.2 Fugitive emission history and regulations

2.2.1 The United States

Problems associated with air pollution were initially acknowledged as an issue of national importance in 1955 in the United States with the passing of the Air Pollution Control Act enacted by Congress. Based on this Act, the US government began providing funding for research and technical assistance related to ways of controlling and mitigating air pollution. The US Clean Air Act followed in 1963 aimed at to decreasing and eventually preventing air pollution. The Clean Air Act is the most influential and modern environmental law and one of the most comprehensive air quality laws in the world. It covers such topics as hazardous air pollutants (HAPs), acid rain, ozone depletion, global warming, etc. Fig. 2.1 illustrates air pollution in Houston in 1964, just 1 year after the advent of the Clean Air Act.

FIGURE 2.1 Air pollution in Houston in 1964. *Shutterstock.*

New amendments were added to the Clean Air Act in 1970 to bring in new approaches, such as standards for measuring and tracking fugitive emission. In the 1980s and 1990s, the Environmental Protection Agency (EPA) started following up with refineries, chemical plants, and other emission-producing industries to assess their leak detection and repair (LDAR) programs to make sure that all the emissions are measured, controlled, and reported. Companies that do not follow the EPA requirements are violating the rules and receive a fine.

The Manufacturers Standardization Society (MSS) of the Valve and Fittings Industry created the first packing test standard in 1997. The first American Petroleum Institute (API) standard for packing tests, API 622, was developed in 2006. ISO 15848-1 was also developed in 2006. The API 624 and 641 standards for valve packing tests were developed in 2014 and 2016, respectively. A detailed description of all of the standards addressing valve fugitive emission is provided in Chapter 4.

2.2.2 The Kyoto protocol

The Kyoto Protocol was initiated in December 1997 in Kyoto, Japan (see Fig. 2.2), and came into force in 2005. The Kyoto Protocol is a United Nations framework convention on climate change that commits industrialized countries and economies in transition to put a limitation on greenhouse gasses. Six major greenhouse emissions should be reduced as per this protocol: carbon dioxide (CO_2), methane (CO_4), nitrous oxide (N_2O), hydrofluorocarbons (HFCs), perfluorocarbons (PFCs), and sulfur hexafluoride (SF_6). Annex B of the Kyoto Protocol sets emission reduction targets for 37 industrialized countries and economies in transition as well as those inside the European Union (EU). The overall aim of this protocol was to reduce emissions by 5% in the period between 2008 and 2012 in comparison with 1990. An amendment was added to the Kyoto Protocol in 2012 in Doha, Qatar, and implemented from 2013 to 2020.

FIGURE 2.2 Kyoto, the city where the Kyoto Protocol was initiated. *Courtesy: Shutterstock.*

2.2.3 TA Luft/VDI 2440

Fugitive emissions in Germany are controlled by TA Luft regulations, which were introduced in 1964. TA Luft is the abbreviation of *Technische Anleitung Reinhaltung der Luft;* its regulations cover air quality requirements, including those for fugitive emissions. TA Luft is based on the Federal Air Pollution Control Act, introduced and enforced by the German Federal Ministry of the Environment. The TA Luft requirements and regulations are also applied in some other European countries. The most recent revisions of TA Luft were created in 2002 with more restrictive fugitive emission requirements on compounds such as sulfur and nitrogen oxides. TA Luft/VDI 2440 is a regulation or directive, and it was the only guideline for preventing fugitive emissions from industrial valves until approximately 2006. This guideline for valve leakage is very general, however, it provides only allowable leakage to the environment without reference to any specific circumstances. TA Luft/VDI 2440 (see Fig. 2.3) is explained in more detail in Chapter 4.

FIGURE 2.3 TA Luft/VDI 2440.

2.3 Fugitive emission sources

Fugitive emission typically occurs in different industries such as oil and gas, chemical, automotive, etc., as per Fig. 2.4. The concentration of this book is on fugitive emission from oil and gas and chemical plants such as refineries and petrochemical plants. Fig. 2.5 illustrates a refinery with different facilities and large amounts of piping. There are many connections in the form of flanged or welded joints between lengths of pipe, and welding joining piping to equipment. Each of these connections could be a source of

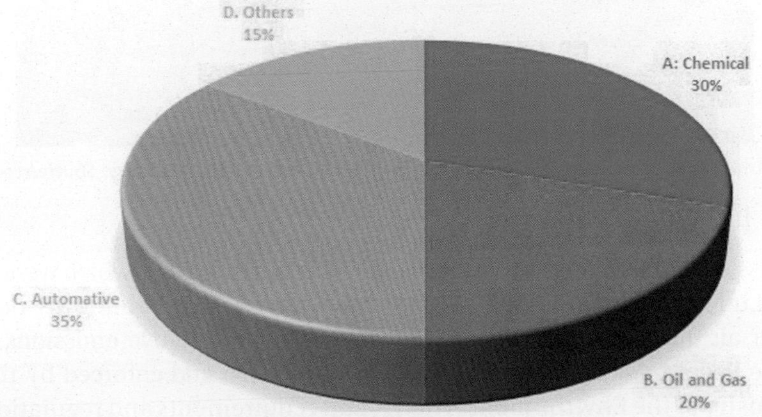

FIGURE 2.4 Fugitive emission percentage from different industry sectors. *The new chart by me.*

FIGURE 2.5 A refinery including facilities and piping. *Courtesy: Shutterstock.*

fugitive emissions, especially volatile organic compounds (VOCs) or HAPs such as methane. VOC and HAP are categorized as gases and compounds that are harmful to the atmosphere. The leakage of oxygen or nitrogen from piping and valves is not considered a fugitive emission, since oxygen and nitrogen already exist in the atmosphere and do not cause any environment pollution. Air pollutants come in two forms: gases and nongaseous compounds. Nongases are liquids and solids such as dust, smoke, mist, etc. Gases include substances such as carbon monoxide, sulfur dioxide, and VOCs.

The accumulation of harmful compounds from different sources of emission or leakage can release a great amount of pollution into the environment. A typical refinery or chemical plant can release around 600–700 tonnes of VOCs per year from leaking equipment and components such as valves, flanges, pumps, etc. According to some studies, oil refineries release approximately 246,069 tonnes (492 million pounds) of VOCs and HAPs each year. The EPA has estimated that oil refineries emit approximately 35 million pounds of methane each year. Thus, one can estimate that almost 7% of VOC emissions from refineries comes from methane.

Research conducted by the University of British Columbia, Vancouver, found that valves, excluding relief valves, are responsible for approximately 60% of the fugitive emissions from refineries and chemical plants. Therefore, industrial valves are essential sources of emissions. In fact, when relief valves are included, 65% of the fugitive emissions from refineries and chemical plants are associated with industrial valves. Two main important fugitive emission challenges are associated with pressure relief valves. The first is that pressure relief valves do not have any stem and the packing box unlike gate, globe, and ball valves which are addressed in the valve fugitive emission standards. The second is that there is no standard or worldwide industry practice for addressing the fugitive emission for pressure relief valves. When it comes to fugitive emissions from valves in refineries and chemical plants, 80% of the emissions are related to leakage from the valve stems and 20% of the leakage comes from other joints. It should be noted that approximately 10,000 to 15,000 valves may exist in a typical refinery or chemical plant. Some studies have proven that the most problematic valves that cause fugitive emission are old valves that were manufactured years before the advent of modern knowledge about valve fugitive emission tests. Thus, fugitive emissions from valves are a key topic taken very seriously by valve suppliers, contractors, and end users in the industry. Fig. 2.6 provides a general overview of the fugitive emission percentages from different facilities and components in a refinery. It is important to note that fugitive emissions are not limited to refineries and can occur in power plants as well as chemical plants. The focus of this book is more on valve stem leakage, which is responsible for approximately half of the emissions that occur inside refineries and other types of oil and gas plants. Statistics show that methane emission alone in the US oil and gas industry in 2018 was equivalent to 175 million tonnes of CO_2 emission. In general, fugitive emission from oil and gas activities can occur as a result of the following:

- Leakage from components, equipment, or joints;
- Vent or drain process;
- Losses due to evaporation;
- Disposal of waste gas sent to a flare system;
- Accidents and equipment failure.

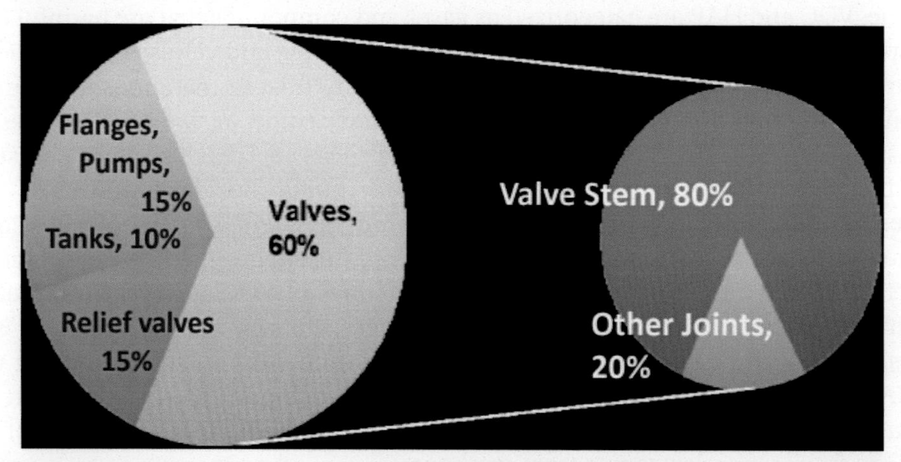

FIGURE 2.6 Fugitive emission percentage from different components in refineries in the oil and gas industry. *Courtesy: Valve World Magazine.*

The flaring of natural gas is one of the major sources of emission to the environment in the oil and gas industry (see Fig. 2.7). Gas flaring is defined as burning gas without using the heat produced from burning. The end user or operator company uses flaring to get rid of the gases extracted from the oil, but the practice releases a large quantity of VOCs, including carbon dioxide, to the environment. It has been reported that the amount of gas flared in 2018 was 3% of the total production of natural gas for that year worldwide. This amount of flared gas to the environment can create high amount of fugitive emission. In general, oil can contain volatile gases such as methane and other types of gas. These gases are separated from the oil in the petroleum industry inside a separator or other type of pressure vessel. The produced gas is often considered to be a

FIGURE 2.7 Gas flare and air pollution in a refinery.

noneconomic amount to sell to a gas buyer, especially when the gas consumer is far from the production site, so the gas is burnt in a flare. However, a number of solutions exist to sell the gas rather than burning it; these include transporting the gas through a pipeline or liquefying it as LNG (liquefied natural gas) and selling it. Another solution is to use the gas instead of burning it; an applicable solution for the upstream sector of the oil and gas industry is to inject the gas into the oil reservoir for advanced oil recovery to improve the oil production rate. The other approach is to use the gas to generate electricity in the turbines rather than flaring it.

2.4 Fugitive emission evaluation and challenges

The evaluation of fugitive emission is not only connected to environmental protection; it is also an economic and political issue. Fugitive emission means losing the source of energy which is costly and environmental pollution due to emission is the concern of many governments. The gas industry produces more emissions than the oil industry. Although the combustion of gas produces carbon dioxide, it is just half of the amount produced by coal and 30% less than the carbon dioxide produced by oil. However, a large amount of methane is released from the gas, which has much more impact on global warming compared to carbon dioxide. Nonetheless, it is noticeable that shifting from coal to gas production has represented a gain and has lowered fugitive emissions for the climate. One of the main challenges is the evaluation and measurement of the fugitive emission amount to the environment. Although there are many practices and methodologies regarding the fugitive emission test, the first and most important challenge is related to the detection of the gas. The detection of VOCs is challenging for several reasons, summarized as follows:

- The emission occurs unintentionally without any control;
- Many VOCs are invisible, odorless, and lighter than air, meaning that they will rise up or diffuse;

- Some of the emissions could occur after the field production life. As an example, some wells may leak after being abandoned.

Significant work is still needed to find approaches to evaluate the fugitive emission amount. Many research groups, governmental organizations, and manufacturers are working to find solutions for the more accurate detection of fugitive emission compounds in the environment.

The second challenge is related to the conversion of gas emissions, such as methane, to the equivalent amount of carbon dioxide. It is important to express the climate impact of those two different emissions in terms of a single unit. The main challenge and question here is to determine how many tonnes of carbon dioxide emit the same amount of pollutants as those emitted from 1 tonne of methane. Early reports indicated that 1 tonne of methane has the same effect in terms of environmental pollution as 21 tonnes of carbon dioxide. The ratio was later increased to 1 tonne of methane equal to 25 tonnes of carbon dioxide, and later to a ratio of 1–28.

2.5 Negative impacts of fugitive emission

The negative impacts and consequences of fugitive emission are significant and include:

1. Environmental pollution and damage, including the negative effects of pollution on human life and health, is the most serious negative impact of fugitive emission. The emission of hydrocarbons and other substances to the environment causes air pollution and contributes to the greenhouse effect. The ***greenhouse effect*** is a process in which the gases in the earth's atmosphere trap the sun's heat. This process causes the earth to warm much more than it should. In fact, ***global warming*** (see Fig. 2.8) is one of the primary and potentially devastating negative consequences of the greenhouse effect. Particular gases that contribute to the greenhouse effect and global warming are called "***greenhouse gases***"; these include carbon dioxide (CO_2), methane (CH_4), etc. Global warming has many negative impacts on the environment, such as melting the glaciers, increasing the incidence of draught, fire, hurricane, and rain storms, causing the rise of seawater, etc., which are illustrated in Fig. 2.9.

   The effect of VOC and HAP on the environment goes beyond the greenhouse effect and global warming. Other environmental effects of VOC and HAP emissions include: photochemical ozone creation potential (POCP) and ozone depletion potential (ODP), as well as toxicity, carcinogenicity, and local nuisance from odor. Ozone layer depletion increases the amount of ultraviolet (UV) radiation that reaches the earth's surface; this radiation can cause skin cancer and other serious health problems. Ground-level ozone creation occurs when the emissions from cars, refineries, plants, and other industrial units react chemically in the presence

FIGURE 2.8 Greenhouse effect and global warming.

FIGURE 2.9 Effects of global warming. *Courtesy: Shutterstock.*

of sunlight. This ozone can reach unhealthy levels, especially during warm days. Inhaling ozone can damage the lungs and cause chest pain, coughing, shortness of breath, and throat irritation. Fig. 2.10 illustrates the green color of the air in the vicinity of a refinery located in Pasadena, California.

FIGURE 2.10 Greenhouse effect and green colored air close to a refinery in Pasadena, California. *Courtesy: Shutterstock.*

2. Spillage of the produced hydrocarbons in the form of oil and gas, which means loss of production. Hydrocarbons are flammable and can cause fire and explosion, leading to loss of asset, loss or decrease of production, damage to human tissues, and loss of human and animal life.
3. Some fugitive emissions contain toxic compounds such as hydrogen sulfide (H_2S). Hydrogen sulfide is a colorless chemical compound with the odor of a rotten egg. Hydrogen sulfide is an undesirable by-product of oil and gas and is extremely toxic, corrosive, and flammable. A small concentration of H_2S can lead to loss of consciousness for a human being. A concentration of a 100 parts per million (ppm) and higher can kill humans very quickly. The other issue associated with this gas is that it has less density compared to air, so it remains at a low level instead of moving upward. Dealing with a type of hydrocarbon that includes hydrogen sulfide necessitates more serious considerations in terms of fugitive emissions. The next section discusses different fluid services according to ASME B31.3 process piping code, and NORSOK M-601 standard for piping and valves.
4. Fines are levied on companies that produce a high amount of fugitive emissions. In fact, some states and local regulatory boards charge annual fees based on the total emissions from facilities and equipment. In 2003, Chevron had to pay USD 3.5 million for fugitive emissions from its refineries; the company spent more than four million for further emission control and other environmental projects. Dupont paid USD 10.25 million in one of the largest penalties due to violation of the EPA's fugitive emission regulations.
5. Damage to the asset: This is not a consequence or impact of the fugitive emission. In fact, damage to the asset could be a reason for fugitive emission initiation.

Fugitive emissions from the valve stem or other joints are due to malfunction and damage to the sealing and packing of the valve, which requires maintenance to repair. Implementation of maintenance on such valves is costly. The cost of maintenance includes spending extra money to pay personnel to do the work, and the cost of replacing or repairing the damaged part. The cost of maintenance could increase over time if the leak is not detected.

In summary, fugitive emission is a hazardous phenomenon that leads to atmospheric pollution and economic loss for oil and gas plants. End users, engineers, and managers of oil and gas plants have many reasons to implement fugitive emission reductions; these include ensuring the safety of the people who work in the plants and those who live nearby; avoiding fines by complying with the rules and regulations, including those that are strict about air pollution; optimizing the plant's energy production; and maximizing the plant's safety and reliability.

2.6 Valve emission sources

2.6.1 Identifying valve leak points

Since industrial valves are the most important sources of emission and leakage in industrial plants, the main focus of this section is valve fugitive emissions. Fig. 2.11 illustrates a control valve that is used in the industry for fluid control or regulation. The valve has an actuator on the top to regulate the motion of the valve. The possible leakage

FIGURE 2.11 Possible leakage points from a control valve. *Courtesy: Shutterstock.*

points from the valve are highlighted in black. The valve has body flanges from both ends that are connected to piping flanges. The flange connections on both ends of the valve are one potential leakage point. The valve has a bonnet or cover that is bolted from the top to the body of the valve. The body and bonnet connection is the second possible leakage point from the valve. The third and final possible leakage point from the valve is through the valve stem packing area. The first two possible leakage points are called static joints because no motion or rotation occurs in these two joints and both are sealed with gaskets. However, the valve packing and stem seal area are dynamic sealing, since the stem that is connected to these seals moves upward and downward. The chance of leakage from dynamic seals is much higher than from static seals. That is the reason why Fig. 2.6 shows that 80% of the leakage from industrial valves is associated with the valve stem or valve stem seals since the seals for the valve stem are dynamic. In fact, the constant movement of the stem upward and downward can wear the packing significantly over time and cause leakage from the valve. The important point is that the movement of the stem is not the only reason leakage could occur from the valve stem packing. Temperature fluctuation, which can lead to expansion or shrinkage of the valve components, including the stem and surrounding packing, could also lead to leakage from the valve. Packing leakage can also occur due to particles in the fluid that damage the packing.

Factors that cause leakage from the flange end connections used to connect the valve to the piping or pipeline system include the following:

1. ***Inappropriate bolt stress:*** Two-mating flanges are connected with bolting, and a gasket is installed between the flanges to provide sealing. As illustrated in Fig. 2.12, one side of the flange is welded to the piping, and the other side of the mating flange is connected to the body of the valve. The required bolt torque values for tightening the flange in order to prevent leakage are given in the installation procedure for the flange. If the provided bolt torques are less than the values given in the procedure, the gasket will not get a sufficient load for sealing and the fluid will leak from the area between the two-mating flanges where the gasket is located. Conversely, overtorqueing the bolts can crash the flange as well as the mating gasket.
2. ***Inadequate thread engagement between bolt and nut:*** Flange bolts and nuts are threaded together. The NORSOK L-004 standard for piping fabrication, installation, flushing, and testing allows the flange bolt threads to be fully extended through the nuts with a minimum of one and a maximum of five threads outside the nut. However, ASME B31.3 allows one thread disengagement between the bolt and nut. As an example, Fig. 2.13 illustrates five threads of bolt out of the nut, which satisfies both the ASME and NOROSK requirements. Thread engagement is important, since it has a direct impact on the engagement between the two-mating flanged and/or flange-ended joints.

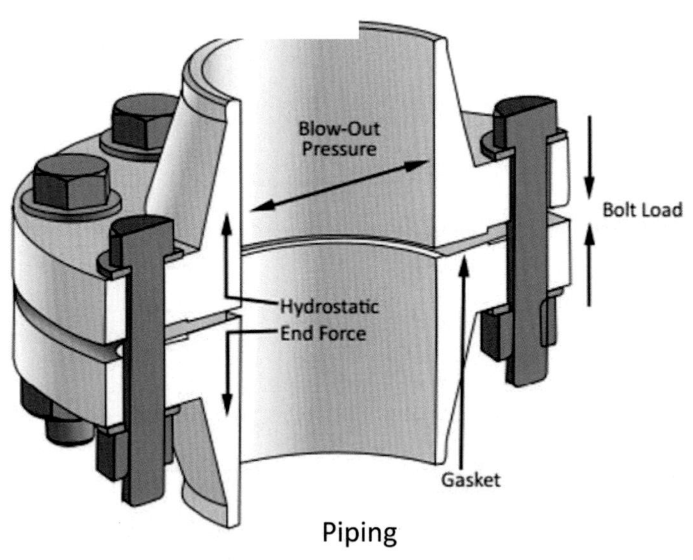

FIGURE 2.12 Two flange connections.

FIGURE 2.13 Bolt and nut thread engagement with five threads out of the nut. *Courtesy: Shutterstock.*

3. ***Improper flange alignment:*** Improper flange alignment causes uneven gasket compression force and results in leakage. The maximum misalignment of the flange in terms of parameter "T" is given as equal to ±1.5 as per Fig. 2.14, extracted from NORSOK L-004.

   Each flange has a specific number of bolts for tightening, which is the nominator of four. The bolting holes of a flange should be aligned with each other. Fig. 2.15 illustrates a flange with eight bolt holes; the maximum misalignment of the bolts in a flange is 1.5 mm. The tolerance of a 1.5 mm maximum misalignment from the flange bolts is taken from the NORSOK L-004 standard.

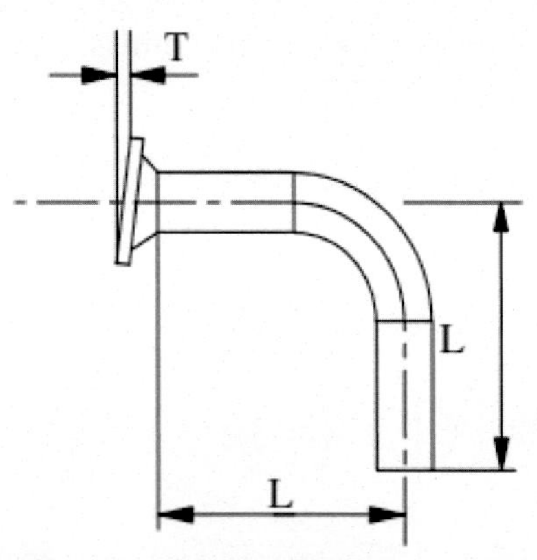

FIGURE 2.14 Maximum misalignment of flange, parameter T = maximum ±1.5. (*NORSOK L-004*).

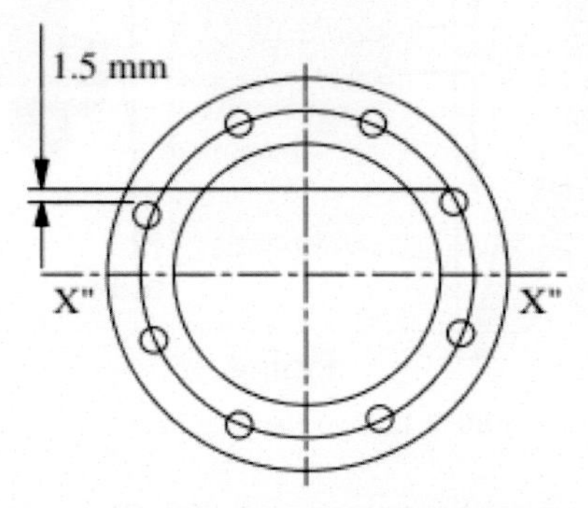

FIGURE 2.15 Maximum misalignment of flange bolts as per NORSOK L-004.

4. ***Improper machining of flange face:*** The face of the flanges should be machined properly as per a flange manufacturing standard such as ASME B16.5 or ASME B16.47 or any equivalent flange standard. Additionally, any damage to the face of the flange or any particle of dirt, welding, etc., could disturb the sealability of the flange. Fig. 2.16 illustrates damage on the flange face where the gasket sits. The flange face should be machined to prevent leakage.

FIGURE 2.16 Damage of the flange face in contact with the gasket. *Courtesy: Shutterstock.*

5. ***Gasket misalignment:*** The gasket should be installed in the center of the flange. Any gasket misalignment can cause leakage. Ring type joint gaskets are installed inside the flange grooves. Spiral wound gaskets are typically installed with a centering ring. Additionally, gasket size and material should be selected properly. Gasket types are explained in Chapter 3.
6. ***Effect of piping stresses and loads:*** Excessive forces and bending movements can loosen the bolting and cause the flange to leak. Some of the factors that result in applying a high amount of load on the joints are lack of piping flexibility, lack of pipe support or improper locations of the support, improper stress analysis job performance, etc. Fig. 2.17 illustrates the effect of excessive loads and bending movement on a flange connection that leads to leakage, deflection of the bolting, and separation of the gasket.

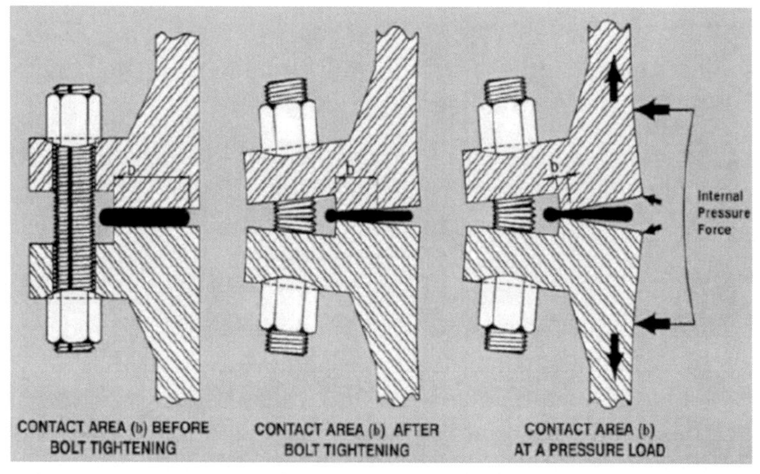

FIGURE 2.17 Excessive piping load on a flange connection and its consequences. *Courtesy: GPI training.*

7. ***Vibration:*** Like stress, vibration can loosen the flange and bolting. Fig. 2.18 illustrates a flange dismantling due to the effect of both stress and vibration.

FIGURE 2.18 Flange joint disconnection due to vibration and stress. *Courtesy: RDI.*

2.6.2 Key factors in low emission solutions

This section provides some of the key factors influential in managing fugitive emissions from industrial valves. They are summarized as follows:

1. ***Location of the plant:*** The quality of oil and gas and the level of impurities found in them vary among oil-producing countries. As an example, there is typically no hydrogen sulfide in the oil and gas produced in Norway. While the quality of the crude may change over time, such that some hydrogen sulfide is produced with the oil, at present, a fugitive emission test is not a concern for valves used in the Norwegian offshore industry. On the other hand, the crude oil produced in countries such as Iran and Kazakhstan in the Caspian Sea typically has a high amount of hydrogen sulfide. Therefore, managing fugitive emissions is stricter and more critical in the oil-producing countries with high levels of impurities and undesirable oil by-products.
2. ***Fluid criticality:*** The different categories of fluid services based on international codes and standards are explained in the next section. Some emissions, such as air or water, are not critical like the emission of oil and gas. Thus, valves that are used for process services such as hydrocarbon are more critical in terms of fugitive emissions compared to those used in utility services such as air and water.
3. ***Location and access to the valves:*** Some of the valves are installed in remote areas with more limited access for leakage monitoring and repair. It is recommended to pay more attention to the packing and sealing design and arrangement of these valves before installation.
4. ***Type of valve:*** In general, the movement of the valve stem could be rotational or linear. Rotational movement of the stem is just 90° for ball valves. Linear movement of the stem is applicable for gate and globe valves. The friction between the stem and the packing and stem seals, as well as the wearing of the stem sealing or packing, is higher for linear stems than for those with a 90-degree rotation.

5. ***Valve direction of installation:*** Valves are installed on either vertical or horizontal lines as per Fig. 2.19. For valves that are installed on the vertical line, the stem of the valve stands horizontally, as illustrated on the left in Fig. 2.20. In that case, extra weight is applied from the stem on the packing area located under the stem and could cause leakage from the packing.

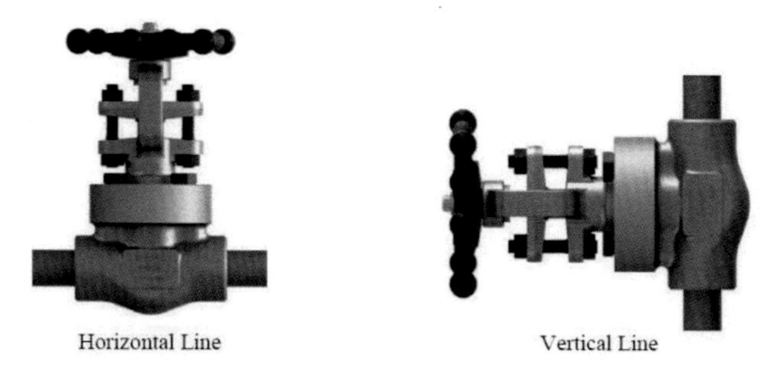

FIGURE 2.19 Valve installation on vertical or horizontal line.

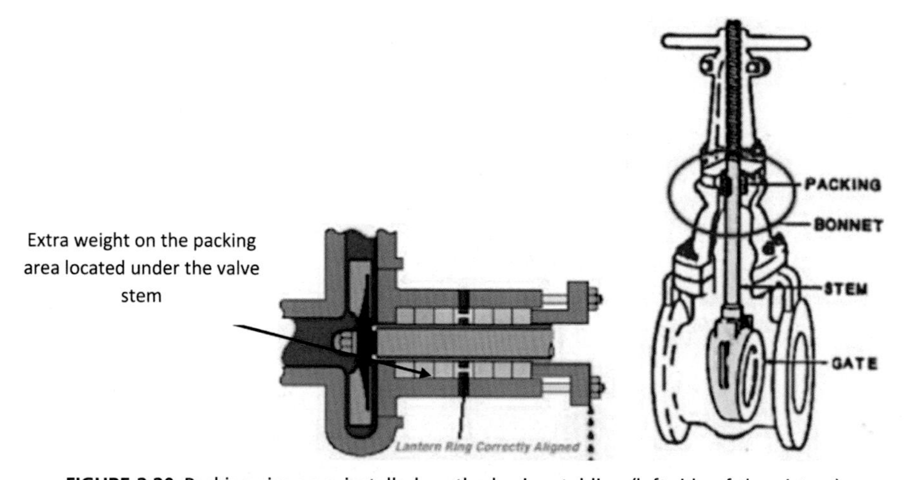

FIGURE 2.20 Packing rings are installed on the horizontal line (left side of the picture).

6. ***Type of valve and piping connections:*** Different factors such as vibration, pulsation, and thermal vibration cause fatigue and eventually leakage in connections such as those found in piping systems, including valves. The difference between thermal and sound vibration is the source of creating the vibration; one is sound and the other one is heat. Different types of connection exist in piping systems; they can be summarized as welding, threaded connections, and tube fitting connections, also called compression fittings. Welded piping and valves can be divided into two categories: buttweld and socket weld. A threaded connection is a weak connection subject to fatigue loads and leakage in corrosive services.

Socket welding is not applicable in the offshore sector of the oil and gas industry. This type of welding is subject to crevice corrosion and is not proposed for piping and valves exposed to fatigue loads. A buttweld connection is resistant to fatigue and vibration and is the most reliable and strongest in general.

7. ***Proper packing, stem design, and alignment:*** Certain essential packing and stem parameters should be considered to achieve a low emission valve. These parameters are proper packing type and material selection, packing internal and external diameter dimensions, tolerances, and proper adjustment of the packing in its place. Valve stems should be manufactured in the correct diameter and installed straight with sufficient and low roughness, as per the relevant valve standard. Excessive stem roughness creates leak points between the packing and the stem. In addition, the amount of bolt torque on the packing rings should be sufficient. Low bolt torque implementation less than the requirement leads to loose packing with a high probability of leakage. Bolt torque higher than what it should be leads to extra friction between the packing and the stem, which makes valve operation difficult. In addition, the friction could damage the packing and cause leakage. Fig. 2.21 illustrates a wedge gate valve and highlights the components whose misalignment leads to leakage and fugitive emission from the valve to the environment.

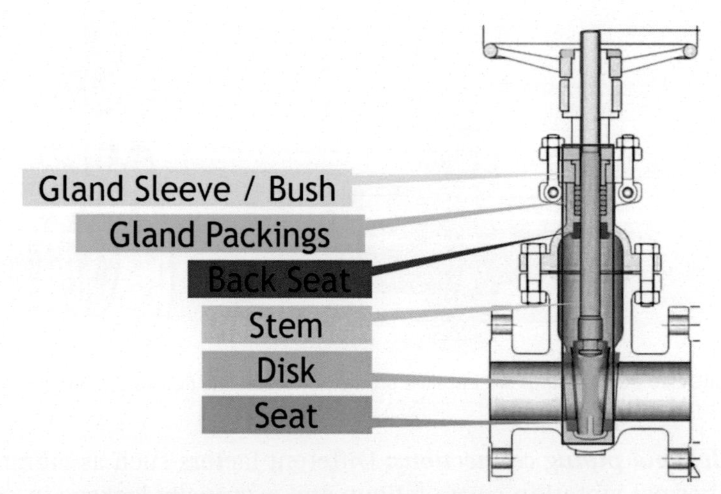

FIGURE 2.21 Wedge gate valve with highlighted components.

8. ***Fugitive emission test:*** Valves, in general, should be tested as per the relevant fugitive emission standards and the end users' specifications or governing laws, which are explained in Chapter 4, to make sure that they achieve low emission compliance.

2.7 Fluid services category

The criticality and significance of fugitive emission control and prevention in piping systems, including valves, is largely dependent on the fluid service. This section provides a short overview of the fluid categories given by ASME B31.3 "Process piping code" as well as the NORSOK M-601 "Welding and inspection of piping" standard.

2.7.1 ASME B31.3

This code addresses process piping and includes three categories of fluid. Each fluid category is connected to a level of hazard based on emission to the environment. Differentiation between fluid services leads to less strict design and testing for less hazardous fluid services, whereas more stringent design and testing are required for more hazardous fluids. Three fluid categories are defined in these codes, and designated as "D," "N," or "M." Each fluid category is defined as follows:

1. ***Category D:*** This category is considered the least hazardous service. The fluids in this category are nonflammable, nontoxic, and not damaging to human tissues. The pressure of this fluid category is a maximum pressure class of 150 equal to 20 bar pressure nominal (PN). In addition, the design temperature of the fluid is between −29°C and 186°C. As an example, a pipe that is handling water in a pressure class of 150 and a design temperature of 20°C is a category D service. Only 5% or no radiography test is required for pipe joints used in service group D.
2. ***Category M:*** This category is considered the most hazardous service. This type of fluid is defined by ASME B31.3 as follows: "a fluid service in which the potential for personnel exposure is judged to be significant and in which a single exposure of a very small quantity of a toxic fluid, caused by leakage, can produce serious irreversible harm to persons on breathing or bodily contact, even when prompt restorative measures are taken." It should be noted that there is no clear list of the fluids in category M. But some fluids, such as hydrogen sulfide, certainly belong to "category M." In order to achieve zero emission from piping systems with category M fluid, a 100% radiography test should be performed on the welding of the pipes to increase the piping joint efficiency to 100% and reduce the chance of leakage to zero.
3. ***Category N:*** If a fluid category does not fall in categories "D" or "M," it is considered category "N" or Normal. This category of fluid is between category "D" and "M" in terms of its hazardous level. Therefore, category "N" fluid is more hazardous compared to category "D" and less hazardous compared to category "M." As an example, piping with water service in pressure class 300 equal to 50 bar can be considered fluid category "N." A 10% radiography test on the piping's circumferential welds should be sufficient for this group.

2.7.2 NORSOK M-601

NORSOK M-601, which has a focus on piping welding and inspection, introduces three fluid categories. Each fluid category corresponds to a nondestructive testing (NDT) group. Table 2.1, extracted from NORSOK M-601, shows the definition of each fluid category. The criticality of the fluid service is indicated by increasing NDT group numbers. The NDT type test becomes more accurate as the NDT group increases. Additionally, the extension of the test is increased by increasing the NDT group. Tables 2.2, extracted from NORSOK M-601, shows the extent of NDT.

Table 2.1 Definition of NDT groups and fluid services as per NORSOK M-601.

NDT group	System service	Pressure rating	Design temp. °C
1[a,b]	Nonflammable and nontoxic fluids only	Class 150 (PN 20)	+29 to 185
2	All systems except those in NDT group 1	Class 150 and class 300 (PN 20 and PN 50)	All
3	All systems	Class 600 and above (≥PN 100)	All

[a]Applicable to carbon steels and stainless steel Type 316 only.
[b]Applicable for all materials in open drain systems.

Table 2.2 NDT extension as per NORSOK M-601.

NDT group	Type of connection	Visual testing, VT%	Volumetric testing, RT%	Surface testing, MT/PT%
1	Buttweld	100	0	0
2	Buttweld	100	10	10
3	Buttweld	100	100	100

2.8 Leak detection and repair

One of the ways to reduce and control emissions in industrial plants is to implement an LDAR program. LDAR is a work practice designed to identify leaking components and equipment and reduce potential emissions by constant control and repair of leaking equipment. A component that is identified as a part of an LDAR program must be monitored at regular intervals to determine whether or not it is leaking. If the component or equipment is found to be leaking during the inspection, it should be repaired or replaced within a specific time frame.

LDAR programs are required by many standards and regulations. The main benefit of LDAR implementation is that it reduces emissions from plants significantly. The EPA estimates that the leakage rate can be reduced by 63% through successful implementation of an LDAR program. Additionally, the EPA predicts that chemical facilities could reduce the number of VOCs by 56% through LDAR. Thus, emission reduction through LDAR has many other benefits, such as reducing production loss, increasing the safety of the workers, and reducing the fees and fines assessed due to fugitive emissions from the plant.

Five steps or elements are defined in LDAR, which are summarized as follows:

1. ***Identifying the components that are at risk of leakage;***
2. ***Definition of leakage;***
3. ***Monitoring components;***
4. ***Leak detection and repair;***
5. ***Recordkeeping.***

2.8.1 Identifying at-risk components

All components and equipment that are subject to leakage should be identified at this stage. As mentioned above, these components could be valves, pumps, and flange connections, at minimum. The best practice is to assign a unique identification number to each of these components to keep track of them. These unique numbers can be indicated on the piping and instrument diagrams (P&IDs) or process flow diagrams (PFDs).

2.8.2 Leak definition

EPA method 21 requires that the leakages from components and facilities subject to leakage be measured in parts per million (ppm). A leak should be detected if it exceeds the leak definition as per the applicable regulation or standard. The definition of leak varies across different standards and regulations. In addition, the type of service in terms of liquid or gas may have a different leak definition or level of maximum allowable leakage. Different regulations define leak in different ppm values, such as 500 ppm, 1000 ppm, and 10,000 ppm. Picking up the best possible leak definition is the key at this stage. The recommendation of best practices for LDAR is to select the lowest and most conservative level of leakage, which is 500 ppm.

2.8.3 Monitoring methods

Different strategies, such as improving the packing and sealing design and arrangement and conducting fugitive emission tests, minimize the risk of fugitive emission during operation. However, there is no guarantee of achieving zero leakage from the valves during operation. The question is, then, what sort of measures should be taken to detect fugitive emissions from valves and actuators during operation? Many people may have an image of manual emission detection by means of a portable device that can be taken around to all the possible leakage points. Indeed, EPA method 21 does propose a procedure to detect VOC leaks from equipment and components by using a portable detecting instrument, as illustrated in Fig. 2.22.

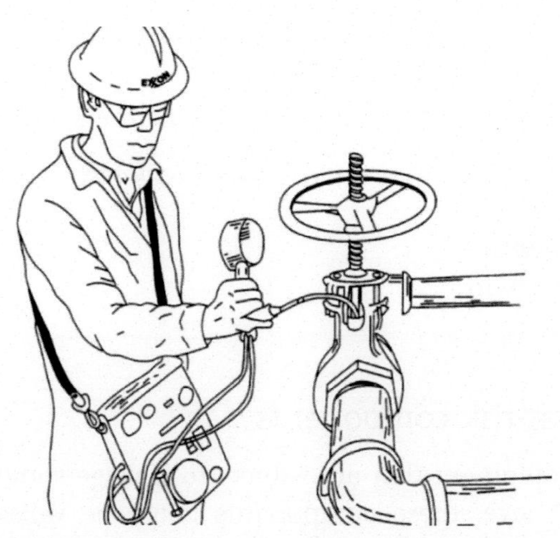

FIGURE 2.22 Detecting leaks from a valve with a portable detecting instrument.

It should be noted that manual monitoring is a time-consuming, inefficient process. Common mistakes and problems that arise when using a portable detecting instrument include not monitoring the leak point long enough, not monitoring all the leak points, reading the leakage amount incorrectly, etc. An alternative way to detect fugitive emissions at an earlier stage before they become critical is to use sensors. Usage of sensors to detect possible fugitive emissions from valves leads to improvement of the safety and reliability of plants during operation. Thus, the best practice for leakage monitoring is to use an automatic, electrical data logger to save time and improve accuracy. Fig. 2.23 illustrates an automatic leak detection method for an oil and gas plant in which the possible leakage points are connected to sensors. The sensors are connected to a leak detection system, and a user sitting in a remote location can monitor the amount of leakage from different components and facilities. Online leak measurement informs the personnel onsite of the detected leakage

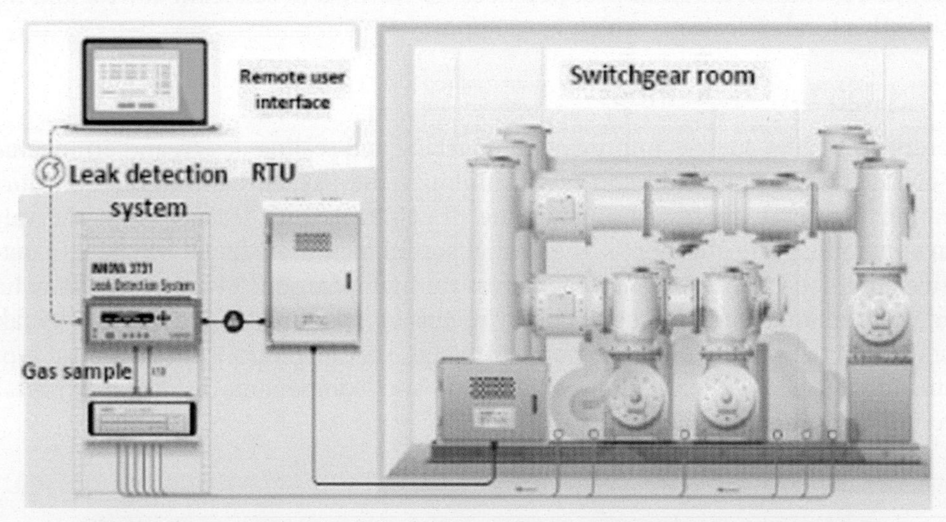

FIGURE 2.23 Automatic online leak detection in an oil and gas plant. *Courtesy: Elsevier.*

FIGURE 2.24 Automatic leak detection sensors on the pipeline by means of acoustic emission. *Courtesy: Shutterstock.*

much faster than the manual leakage monitoring approach. Even if another method of monitoring instead of EPA 21 is used, it is proposed based on LDAR best practice to use the EPA 21 method and compare it with the results provided by the alternative monitoring method. Additionally, frequent monitoring of leak points is proposed. One of the automatic technologies that can be used for measuring leakage in a long onshore or offshore pipeline is the acoustic emission approach. As illustrated in Fig. 2.24, different sensors that are sensitive to noise level are located on a long pipeline. Any leakage from the pipeline increases the level of noise, which can be detected by a sensor located in the vicinity of the leakage and reported to the leak detection system or room.

2.8.4 Repairing components

Leaking components should be repaired as soon as possible and practical, but no later than a specific number of calendar days. Typically, the regulations allow 5 days for the first attempt at the repair and 15 days for the final attempt. Repairing an industrial valve can include tightening the bonnet, body, or gland flange bolts. Changing the packing or injecting lubrication into the packing are other repair activities. The component or equipment is considered to be repaired if the amount of leakage after the repair or maintenance is less than the defined leak in LDAR strategy. Four best practices are provided for repair as follows:

- Preparing a timeline or schedule for repairing components;
- Attempting to repair the leak as soon as possible;
- Monitoring the repaired components or facilities after the repair for several days to make sure that the leak has been successfully repaired;

- Replacing the component if repair is not possible;
- Replacing major leak points with leakless solutions or technologies.

2.8.5 Recordkeeping

It is essential to maintain a record of the components, such as valves, that have undergone a repair due to leakage. In addition, it is important to keep a record of all leak monitoring and repair activities, including the dates and results. It is proposed by the best practice of LDAR to attach an identification tag to equipment and components that have leaked. The best practice is to perform both an internal and a third-party audit on the LDAR records. The records should be updated regularly. Fig. 2.25 illustrates how the implementation of the abovementioned steps through LDAR reduces fugitive emissions significantly.

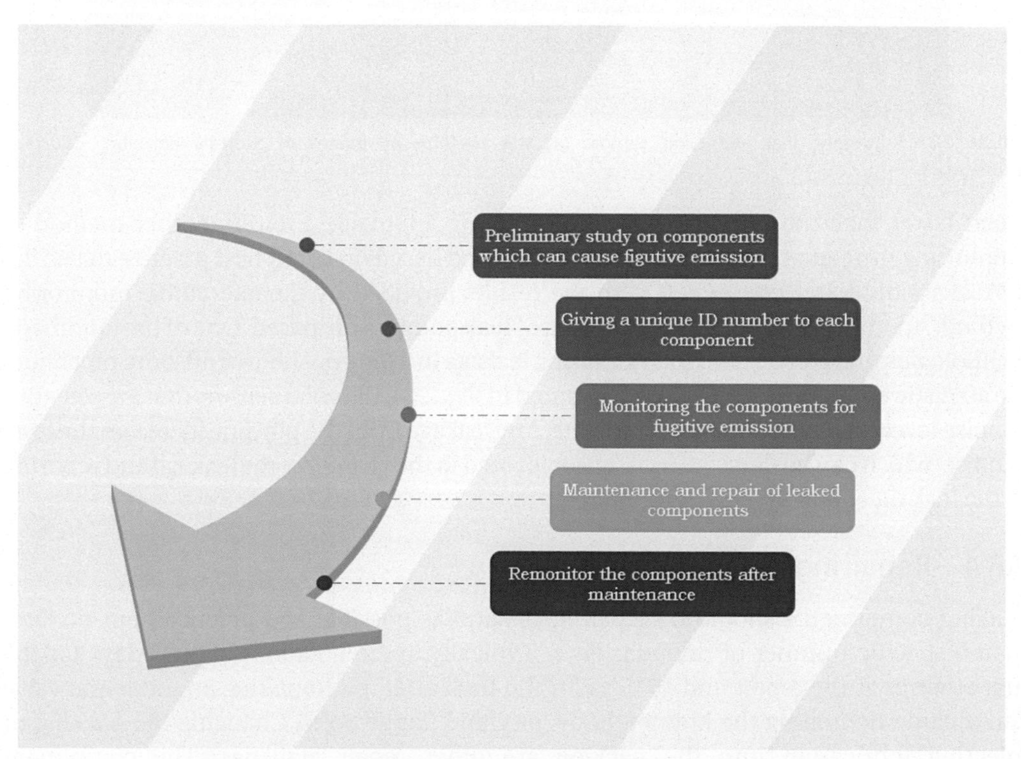

FIGURE 2.25 Fugitive emission reduction through LDAR implementation.

2.9 Questions and answers

1. Which sentence is not correct regarding fugitive emission?
   A. Valves are the main component from which fugitive emission takes place.
   B. Fugitive emission in chemical plants is typically higher than in the oil and gas sector.

C. Global warming and the greenhouse effect are two negative, interrelated consequences of fugitive emission.

D. Industrial valves, including safety valves, are responsible for more than 75% of the emissions from oil and gas refineries.

Answer: Options A, B, and C are correct. However, option D is not correct since, as per Fig. 2.6, industrial valves, including safety valves, are responsible for approximately 75% of emissions from refineries.

2. Which sentence is correct about fugitive emission?

A. There is an accurate and easy way to convert methane emission to the equivalent carbon dioxide emission.

B. Flaring gas is not a major concern in terms of fugitive emission and air pollution.

C. The major leaks from valves come from the body and bonnet joints.

D. Category M fluid, as per ASME B31.3, is the most dangerous fluid.

Answer: Option A is not correct; it is a challenging task to convert methane emission to the equivalent carbon dioxide emission, since there is no accurate and widely accepted approach for this conversation. Option B is not correct since flaring gas into the environment is a major fugitive emission concern. Option C is not correct, since most of the leakage from valves originates from the valve stem and not the body and bonnet. Option D is correct, since category M is the most dangerous type of fluid defined in ASME B31.3, and has extremely negative effects on human tissues and the environment.

3. Which solution is mainly used in the upstream sector of the oil and gas industry to utilize the gas instead of flaring it, thus reducing fugitive emissions?

A. Liquefy the gas and transport it through gas pipelines

B. Inject the gas into the oil reservoir

C. Use the gas to fuel turbines to generate electricity

D. Inject the gas into the ocean

Answer: Liquefying the gas and transporting it through a pipeline is mainly done in the downstream sector of the oil and gas industry, so option A is not correct. Option B is correct, since the gas is mainly injected into the reservoir for advanced oil recovery in the upstream sector of the oil and gas industry. Using the gas for turbines is not a solution in the upstream sector. Injecting the gas into the ocean is not a practice.

4. Which sentence is not correct regarding fugitive emission control and measurement?

A. LDAR implementation in industrial plants can reduce fugitive emissions significantly.

B. One of the main challenges in fugitive emission control is detecting the source of emission.

C. Carbon dioxide and methane are two fugitive emission compounds.

D. Carbon dioxide emission is more dangerous than the same weight/amount of methane emission.

Answer: Option A is correct, since LDAR can reduce fugitive emission significantly more than 50%. Option B is also correct; the detection of fugitive emission gases is challenging, since they are volatile and have no odor. Option C is also correct, since both methane and carbon dioxide are fugitive emission compounds. Option D is wrong, since methane releases more air pollution and carbon emission compared to carbon dioxide of the same weight.

5. Which sentences are correct regarding the steps of LDAR?
   A. The best practice is to define the leakage for the components between 500 and 10,000 ppm, as achieving a leak rate of less than 500 ppm is not practical.
   B. Using an online leak monitoring and detection system completely removes the necessity for using a portable leak detecting instrument.
   C. It is the best practice to identify and provide tags on leakage points and components.
   D. Repairing a leaking component within 10 days is considered a very fast repair attempt.

   Answer: The best practice for defining allowable leakage is to consider the minimum amount given in the rules and regulations, which is 500 ppm. Thus, option A is not correct. Using an online monitoring does not remove the need for a portable leak detecting instrument, so option B is not correct. Option C is correct; the best practice is to identify and provide tag numbers for components subject to leakage as the first step of LDAR implementation. Option D is not correct, since 5 days is considered the fastest attempt to repair a component or piece of equipment.

6. Fugitive emission has negative impacts on health, safety, and environment (HSE) issues. Which negative impacts of fugitive emission listed below have no impact on HSE?
   A. The greenhouse effect and global warming
   B. Loss of human life due to hydrogen sulfide in the methane leaked from a valve
   C. Loss of asset and fines levied on companies due to fugitive emission
   D. Possible fire from flammable emissions

   Answer: All of the options except for option C address the negative effects of fugitive emission on HSE. Option C addresses the negative economic impacts of fugitive emission, so it is the correct answer.

7. Which sentence is not correct?
   A. The requirement for NDT implementation on fluid services is always stricter in NORSOK M-601 than ASME B31.3.
   B. Leakage is more likely from a valve when it is installed vertically with the stem on the horizontal plane.
   C. Buttweld connections in piping and valves are stronger than socket weld connections.
   D. More attention should be paid to fugitive emission from valves when they are installed in remote areas.

Answer: Option A is not correct. NORSOK is not always more stringent compared to ASME B31.3 when it comes to NDT percentage for fluid categories. As an example, a toxic fluid with a pressure class of 300 is categorized as NDT group 2 with a 10% radiography test on the circumferential joints as per NORSOK M-610 standard. However, the same fluid is categorized as fluid group M with a 100% radiography test on the circumferential piping weld joints as per ASME B31.3 standard. Thus, the requirements of the NDT test extension in this case is more stringent in ASME B31.3 than in NORSOK M-610. Option B is correct, since installing the valve on the vertical line with the stem on the horizontal plane puts more pressure from the stem on the packing area located under the stem and can cause leakage; this risk of leakage is not possible for a valve installed a horizontal line with a vertically installed stem. Option C is correct; a socket weld is not a very robust welding connection. It is common for small piping, typically in sizes of 2″ and less. In addition, socket weld connections are at risk of crevice corrosion. Option D is correct; more attention regarding fugitive emission should be paid to valves installed in remote areas.

8. Which of the factors below are not causes of leakage from a flange?

A. Low strength bolting material in high-pressure class piping and poor gasket material selection

B. Supporting the piping system and providing flexibility to the pipe

C. Improper flange face roughness

D. Improper flange joints and bolt hole alignments

Answer: According to ASME B31.3, which addresses process piping requirements, the bolting for high-pressure class piping such as class 600 equal to 100 bar nominal pressure should be high-yield bolting with a yield strength of more than 30 ksi equal to 30,000 psi. Thus, low strength bolting material with less than 30 ksi for a high-pressure piping system could cause leakage from the flange. Poor gasket material selection is another cause of leakage from the piping flange joints. Thus, option A is correct. Option B is not correct, since supporting piping and providing flexibility prevents leakage from piping and flange connections. Options C and D are correct; they are explained in the text.

9. Which statement is correct as per the given standards?

A. Two disengagement of threads between the bolts and nuts in a flange connection is acceptable in both ASME B31.3 and NORSOK L-004.

B. The misalignment of the flange face from the vertical line can be 2 mm as per the NORSOK L-004 standard.

C. A 50% radiography test (nondestructive test) on circumferential piping welds to be used in a service with a high amount of hydrogen sulfide concentration is sufficient according to ASME B31.3, process piping code.

D. The machining roughness for the surfaces of flanges used in high-pressure piping classes should be less than the roughness of the surface of flanges in low-pressure piping classes according to the ASME B16.5 flange standard.

Answer: Two thread disengagement between bolts and nuts is not acceptable in either ASME B31.3 or NORSOK L-004, so option A is not correct. The maximum misalignment of the flange face from the vertical line is 1.5 mm (refer to Fig. 2.14), so option B is not correct. The percentage of radiography test on the longitudinal welds for a toxic service containing a high amount of hydrogen sulfide as per ASME B31.3 is 100%, so option C is not correct. Option D is correct since the maximum roughness of a ring type joint flange face used for high-pressure class piping should be 63 μin as per ASME B16.5. However, other flange faces, such as raised face and flat face, typically have 125–250 μin. In fact, less roughness is required for the tight sealing of ring type joint metallic gaskets in terms of the metal-to-metal contact between the gasket and the flange face.

10. Which item could not affect leakage from valve stem seals?

A. Type of stem movement (linear or rotary)
B. Elevation of the installed valve
C. Type of fluid (gas, liquid)
D. Number of packing rings

Answer: The type of stem movement could affect leakage from the valve stem. Generally, valves with linear stem movement produce more leakage from the stem, since the friction between the stem and the packing or stem sealing is greater in valves with linear stem movement. Therefore, Option A is correct. Option B is not correct, since the elevation of the valve installation does not affect the stem sealing capability of the valve. The type of fluid affects the amount of leakage from the stem packing. Typically, gases have a higher tendency to escape from the stem packing than do liquids. Thus, option C is correct. Option D is correct, since higher numbers of packing rings could reduce the chance of leakage from the stem area.

Bibliography

[1] American Society of Mechanical Engineers (ASME) B16.5. Pipe flanges and flanged fittings: NPS $^{1}/_{2}$ " through NPS 24 metric/inch standard. 2017 [New York, NY].

[2] American Society of Mechanical Engineers (ASME) B16.47. Pipe flanges and flanged fittings: NPS 26 through NPS 60 metric/inch standard. 2017 [New York, NY].

[3] American Society of Mechanical Engineers (ASME) B31.3. Process piping. 2012 [New York, NY].

[4] Cox JC. Thwarting fluid-system leaks: saving energy. Reducing Leaks 2010;15(9):75–9.

[5] Environment Protection Agency (EPA). Leak detection and repair—a best practices guide. 2007 [Washington DC. USA].

[6] Gielen D, Kram T. The role of Non- CO_2 greenhouse gases in meeting Kyoto targets. Climate change and economical modelling: background analysis for the Kyoto analysis. 2010.

[7] Gobind K. Valve fugitive emissions vs. current industry practices—Part 2. Valve World Mag 2016; 21(10):121–2.

[8] NELES. Fugitive emission efficiency. 2014 [online] Available from: https://www.neles.com/insights/articles/fugitive-emissions-efficiency/. [Accessed 4 December 2020].

[9] NORSOK M-601. Welding and inspection of piping. 5th ed. 2008 [Lysaker, Norway].

[10] NORSOK L-004. Piping fabrication, installation, flushing and testing. 2nd ed. 2010 [Lysaker, Norway].

[11] Laconde T. Fugitive emissions: a blind spot in the fight against climate change. Climate change organization. Climate change-annual report—global observatory on non-state climate action. 2018.

[12] Picard D. Fugitive emission from oil and natural gas activities. J Good Pract Guideline Uncertainty Manage Natl Greenhouse Gas Invent 1999:103–27.

[13] Shoberg A. Everything you need to know about valve fugitive emission reduction. HABONIM Industrial Valves & Actuators; 2020 [online] Available from: https://blog.habonim.com/everything-you-need-to-know-about-valve-fugitive-emission-reduction. [Accessed 27 October 2020].

[14] United Nations Climate Change. What is the Kyoto protocol?. 2020 [online] Available from: https://unfccc.int/kyoto_protocol. [Accessed 16 November 2020].

[15] Winberry J. Introduction to fugitive emissions monitoring. 1st ed. North Carolina. USA: Environmental Programs; 2000.

[16] Wing MS, Smith BK. Controlling and monitoring control valve fugitive emissions. 2012 [online] Available from: http://www.valvemagazine.com/magazine/sections/features/4393-controlling-and-monitoring-control-valve-fugitive-emissions.html. [Accessed 1 November 2020].

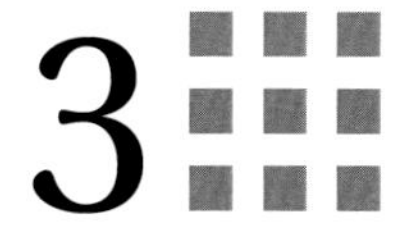

Valve sealing and packing

3.1 Introduction to industrial valve seals

Valve design is generally considered a broad subject that includes various considerations such as material selection, wall thickness calculation, sealing design, design of the pressure-controlling and internal parts of the valve, and design of the pressure-containing parts of the valve such as the body and bonnet, stem, and bolting. Leakage of fluid from the valve to the environment is highly possible for several reasons, such as poor sealing selection, lack of sealing tightness inside the valve, etc. Thus, it can be concluded that valve seals, including packings, are critical parts of industrial valves. Therefore, it is essential to focus on valve seals in a separate chapter due to the importance of minimizing the risk of the emission of gasses and other VOCs through valve seals to the environment. It is essential to select, design, and arrange valve seals in the best way to prevent leakage and increase the life and performance of the valve. The definitions of both seals and packing are provided in Chapter 1. It should be borne in mind that regulating agencies are becoming more and more strict regarding fugitive emissions from valves for the reasons explained in the previous chapter.

There are three main, important improvements connected to proper seal selection and application of industrial valves:

1. Increasing valve life and improving performance;
2. Reducing leakage and fugitive emissions;
3. Reducing maintenance time and associated costs.

3.2 Valve components seals

Several parts of valves require sealing; these are summarized below:

1. ***Stem sealing:*** Stem sealing or packing is the most important seal in industrial valves. The stem sealing is typically called packing if it is made of graphite. Graphite packing is more common for gate and globe valves. O-rings are another type of stem seal. Multiple O-rings around the stem are common in ball valves. Stem seals or packing are important, since any leakage from the seals or packing rings goes directly to the environment from the inside of the valve. Fig. 3.1 illustrates a ball valve and highlights the stem and stem sealing.
2. ***Body and bonnet sealing:*** Some valves, such as gate and top entry ball valves, have a bonnet or cover on the body. Normally, a seal in the form of a gasket is placed

Prevention of Valve Fugitive Emissions in the Oil and Gas Industry. https://doi.org/10.1016/B978-0-323-91862-6.00001-0

FIGURE 3.1 A ball valve with highlighted stem and stem sealing.

between the body and bonnet to prevent leakage from these two components. Fig. 3.2 illustrates a slab gate valve in a high-pressure class, highlighting the body and bonnet and the metallic seal ring or gasket between them.

FIGURE 3.2 A slab gate valve.

3. ***Body pieces sealing:*** Some industrial valves have seals between two body pieces. Fig. 3.3 illustrates a side-entry ball valve with three body pieces connected with bolts and nuts. There are two seals or gaskets between each pair of body pieces.

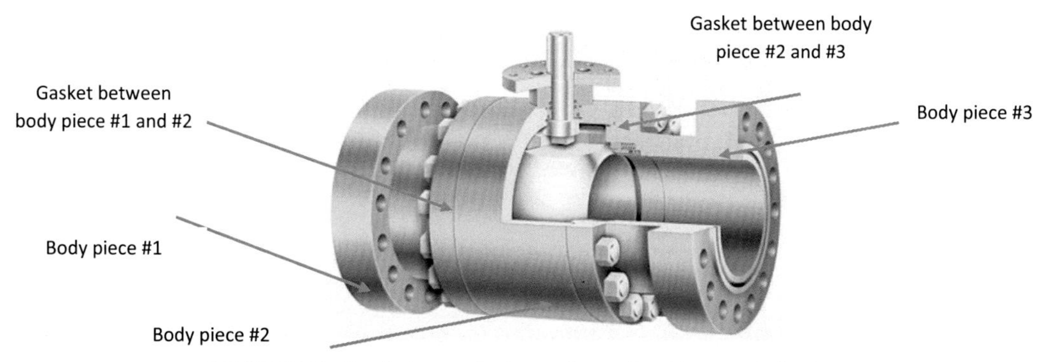

FIGURE 3.3 Sealing between the body pieces of a side-entry ball valve.

4. ***Other valve connection seals:*** Seals exist on other valve connections, such as the connection between the body and bottom cover in a butterfly valve. Butterfly valves are used, like ball valves, to stop and start flow. The advantage of butterfly valves is that they are lighter, more compact, and cheaper than ball valves. Fig. 3.4 illustrates the body of a butterfly valve undergoing a hydrostatic body pressure test. The figure shows some amount of water leakage from the area between the bottom cover and the body of the valve.

FIGURE 3.4 Leakage from the area between the body and bottom cover of a butterfly valve during the hydrostatic pressure test.

Subsequent to the test, the bottom cover of the valve was disassembled from the body and the seal between the bottom cover and the body of the valve was inspected. The seal between these two components is a black O-ring in Viton material; it is damaged, as illustrated in Fig. 3.5. The leakage during the test was a result of the damaged O-ring. O-rings and Viton are defined later in this chapter. A seal could be placed between the

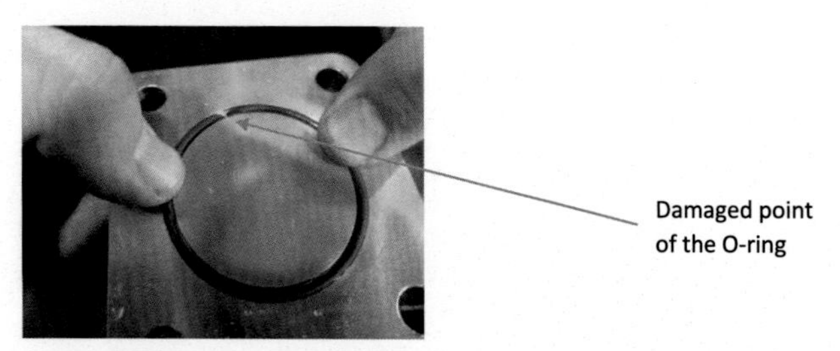

FIGURE 3.5 Damaged Viton O-ring between the bottom cover and body of a valve.

bottom cover and body of a valve in addition to being placed between body pieces, and between the body and bonnet.

Some valves have auxiliary connections, such as a vent and drain plug or flange. Auxiliary connections and their purposes are clearly explained in Chapter 1. Plug connections have threads, which are prone to the risk of leakage during operation. Thus, seals or sealants such as PTFE (Teflon) or Loctite (see Fig. 3.6) are used to secure the plugs in the valve body, ensuring no leakage during operation.

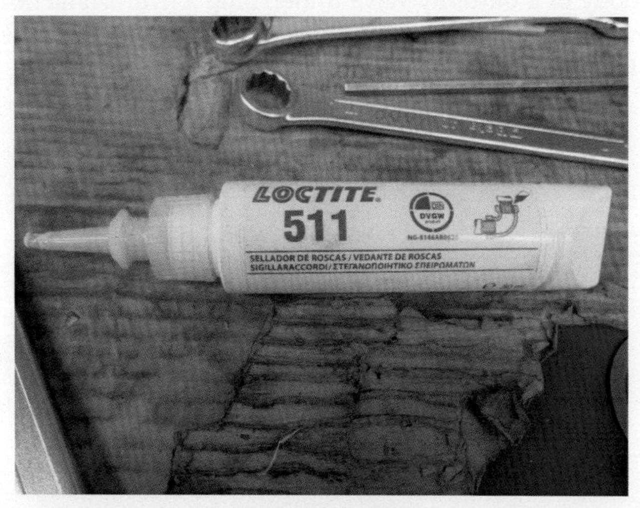

FIGURE 3.6 Loctite.

Internal seals: These seals can be categorized into two groups: soft or metallic. A soft seat seals the area between the valve closure member and the back seat or seat carrier. This soft seat is typically called a seat insert, illustrated in Fig. 3.7. No leakage is allowed for soft seat inserts (refer to different valve testing standards such as API 598 and ISO 5208). Alternatively, the valve could have a metallic seat, without any seat insert, that is in contact with a valve closure member such as a ball. Typically, some amount of leakage, expressed in terms of leakage class, is allowed from a metal seat. Leakage from a seat insert or metallic seat is considered internal leakage that will not

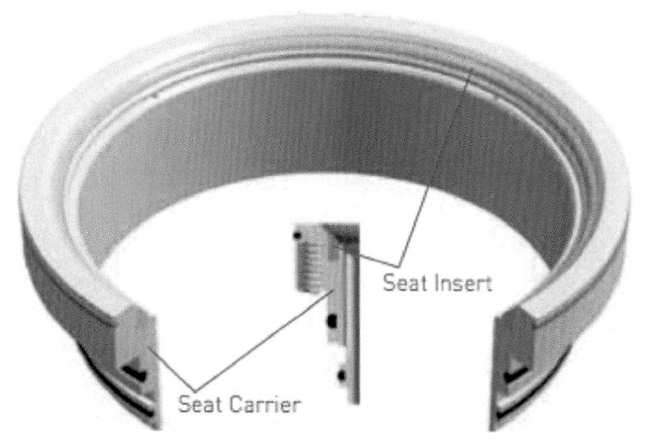

FIGURE 3.7 Seat carrier and seat insert arrangement in a ball valve.

cause emission to the environment. Different factors result in the leakage of fluid from valve seats, such as poor seat insert design, poor material selection for the seat insert and seat carrier, damage due to the particles inside the fluid, etc. Fig. 3.8 illustrates how the seat carrier, which is typically in metallic material, and the soft seat insert are positioned and where they are located in a ball valve. In general, valve seats can be fixed or floating. Floating seats, in which the springs are located behind the seat carrier and push the seat insert toward the ball to prevent leakage, are more common. Internal leakage from the valve can reduce the fluid volume passage through the valve if the valve is in open position; it can also lead to overpressure in the valve cavity and other operational problems. In addition to the sealing between the seats and the closure member such as ball, there are other internal seals in the valve to prevent leakage from the area between the seat and body. These seals are normally made of soft materials, such as graphite or viton or lip seal.

FIGURE 3.8 Positions and locations of seat carrier and seat insert in a ball valve.

3.3 Dynamic and static seals

Seals are divided into two categories: static and dynamic. Static seals are used between two nonmoving parts of the valve, and dynamic seals are used between at least one moving or rotating valve part and another part that may be moving or nonmoving. Dynamic seals can be more critical, since they are at higher risk of wearing and tearing due to friction with one or two moving parts. Based on the above definition, stem seals and the seat insert are categorized as dynamic seals. Stem seals are located between the valve stem and body, and the valve stem is a moving part. The seat insert is located between the valve closure member, such as the ball, and the seat; the valve closure member is a moving part. Seals between the valve body and bonnet or between valve body pieces are categorized as static seals, since both the body and bonnet are nonmoving components. No static or dynamic seal category could be defined for Teflon or Loctite sealant, since these sealing components are not a part of the valve.

3.4 Valve stem seals

When it comes to fugitive emission considerations, valve stem seals are known as the most important leakage points for two reasons. First, valve stem seals are categorized as dynamic seals due to the movement of the stem. Second, unlike the valve seat or seat insert, leakage from the valve stem seals is released to the environment and is a main source of fugitive emission. Thus, it is worth focusing in more detail on valve stem seals, including their materials and arrangements, as well as their effects on valve operation.

3.5 Reasons for valve leakage

In general, three main factors could lead to valve leakage, as summarized below:

1. Wear and tear or corrosion of the valve's metallic components;
2. Faulty connection or misalignment between some of the valve components; and
3. Malfunction of stem seals due to wear and tear, corrosion, or mechanical damage.

3.6 Valve sealing types

Control of fluid loss is essential in the mechanical equipment, including valves, used to handle fluid. Various sealing types are used in valves to prevent leakage from the shaft, body and bonnet, or body sections. The first step in valve seal selection is to decide on the type of seal. Multiple parameters should be taken into account during sealing selection, such as resistance to the operating temperature and pressure. In some cases, the seals are required to withstand very high pressure and a wide range of temperatures from extremely hot to extremely cold. Compatibility of the sealing with the fluid service is another important parameter for sealing selection. In addition, friction between the sealing and the stem is an essential parameter, since high-friction sealing can increase the required force or torque needed for the operation of the valve. The sealing should have sufficient resistance against the loads and cycles to prevent wear and tear. Proper sealability and economic considerations are two other important factors that should be taken into account during sealing selection for industrial valves and other components.

In general, valve seals can be categorized into gaskets, O-rings, lip seals, V-packs, compression packing, and bellows. Each of these sealing categories are explained below.

3.6.1 Gaskets

Gaskets are typically used between the body and bonnet or body pieces of a valve. Gaskets are covered by ASME B16.20 "Metallic gaskets for pipe flanges" and ASME B16.21 "Non-metallic flat gaskets for pipe flanges" for topside valves, and by API 6A "Specification for wellhead and Christmas tree equipment" for subsea valves. In total, three types of gaskets are used for valve sealing, which are explained in the following section. Different factors contribute to gasket malfunction or failure, such as improper gasket type selection, poor gasket material selection, improper installation, lack of lubrication, improper surface roughness, too much hardness or softness of the gasket, inadequate bolt torque on the mating flanges between which the gasket is installed, etc.

3.6.1.1 Nonmetallic flat gaskets

Nonmetallic flat gaskets are made of materials such as aramid fiber, glass fiber, Teflon (PTFE), graphite, or elastomers. Flat gaskets are covered by the ASME B16.21 standard. Fig. 3.9 illustrates a flat gasket made of rubber. Rubber is a type of elastomer, which is defined in Chapter 1. Nonmetallic flat gaskets are typically used in low-pressure classes such as CL150 equal to 20 bar and nonaggressive and nonprocess services such as air and water.

FIGURE 3.9 Flat rubber gasket.

3.6.1.2 Semimetallic gaskets

Semimetallic gaskets are made of both metallic and nonmetallic materials. The most common type of this gasket is called a spiral wound gasket, illustrated in Fig. 3.10. Semimetallic gaskets contain three parts: The first is an inner ring made of a corrosion-resistant alloy such as stainless steel 316 or a nickel alloy such as Inconel 625 used to increase the strength of the gasket against the loads. In some cases, such as small gaskets, the inner ring may be deleted. The second part is an outer ring, which is also a metallic part that could be in carbon steel in onshore plants; this part is used for proper alignment of the gasket. The third part is a winding and filler, which is located in the middle of the inner and outer ring to provide sealing. This part is typically made of a steel and graphite filler. The main function of the filler and winding is to provide sealing. Spiral wound gaskets are covered by the ASME B16.20 standard.

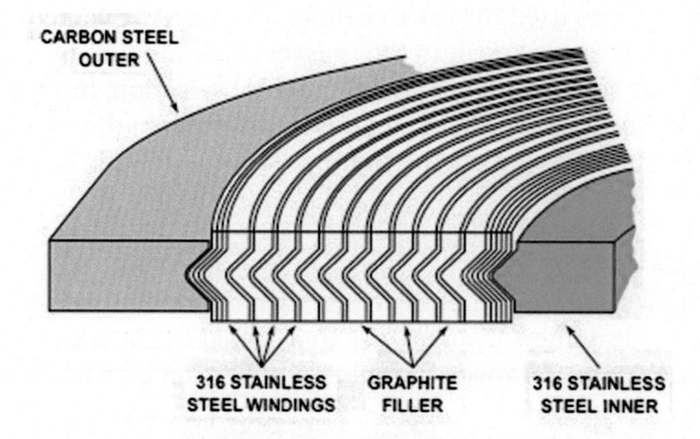

FIGURE 3.10 Spiral wound gasket.

3.6.1.3 Metallic gaskets

Metallic gaskets are mainly called ring-type joint (RTJ) gaskets and are used in high-pressure classes, typically class 600 equal to 100 bar pressure nominal. RTJ gaskets are machined precisely to be placed inside the grooves of mating flanges or other mechanical components. These gaskets are covered by ASME B16.20 for the topside sector of the oil and gas industry, or API 6A for the subsea sector. Fig. 3.11 illustrates an RTJ gasket inside the groove of a flange. Metallic gaskets can be oval or octagonal in shape, as illustrated in Fig. 3.12.

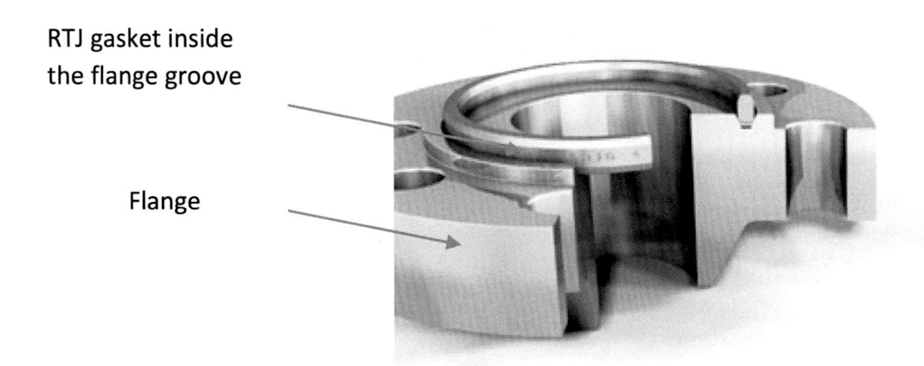

FIGURE 3.11 Ring-type joint gasket inside the groove of a flange.

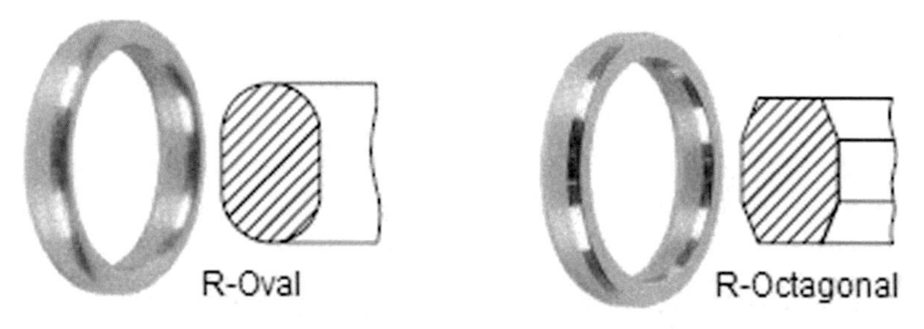

FIGURE 3.12 Oval or octagonal RTJ gaskets.

3.6.2 O-rings

O-rings (see Fig. 3.13) are another type of seal that could be used for stem seals, body and bonnet, or body pieces seals in nonaggressive fluid services, and internal seals for valve components, such as between the body and seat of a valve. O-rings are typically installed in machined grooves. In some cases, when the grooves are in carbon steel, they may be weld overlaid with an exotic material such as Inconel 625 to prevent crevice corrosion under the O-rings. In general, fluid can be trapped under O-rings inside the grooves and cause crevice corrosion. O-rings in industrial valves are used to block the

FIGURE 3.13 Valve O-rings.

path of fluid, which could be gas or liquid or a mix of both, to the inside of the valve and into the environment. O-rings are typically squeezed against the opposite wall or the grooves to maintain perfect sealing at high- and low-pressure conditions.

3.6.3 Lip seals

Lip seals are typically made of a soft material, such as PTFE, energized with a metallic spring in a corrosion-resistant alloy (see Fig. 3.14). The spring provides the required force for sealing the soft material. Lip seals are normally more robust than elastomeric O-rings; they have antiexplosive decompression (AED) by default and can be used for a variety of chemicals as well as temperature and pressure ranges. The important point is that a lip seal can provide sealing in only one direction, so it is not a bidirectional sealing like an O-ring. The fluid can only be sealed with lip seals from the direction where the metallic springs are located or the “lip side.” The temperature range of the lip seal depends on the temperature range of both its metallic and soft materials. A lip seal with a PTFE and Elgiloy spring could be used for a temperature range of approximately −196°C to 250°C, so it can be also used for cryogenic or very cold temperatures.

FIGURE 3.14 Spring energized lip seal.

3.6.4 V-packs

The V-pack illustrated in Fig. 3.15, also called Chevron V-pack sealing, includes male and female adaptors or layers, which could be a group of three or five layers or even more. This type of sealing can withstand very high pressure and can be used for heavy-duty applications like the valve stem sealing or body and closure member sealing of subsea valves. V-packs are also suitable sealing for rods and pistons, which are not applicable to valves. Fig. 3.15 shows a V-pack made of PEEK and PTFE thermoplastic seals. V-packs

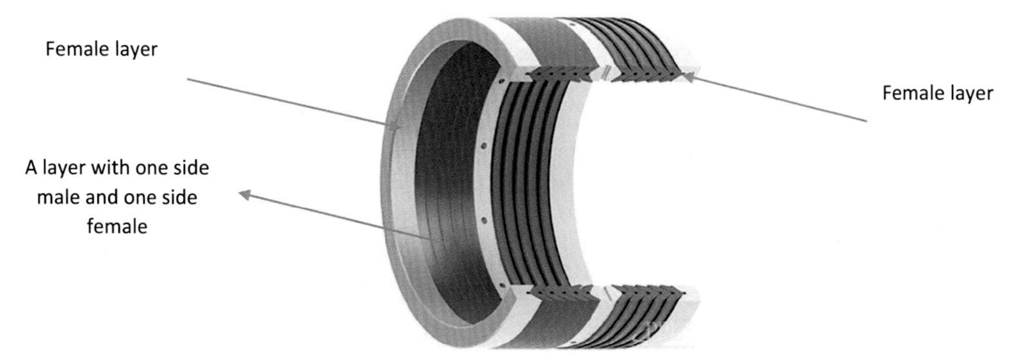

FIGURE 3.15 V-pack with male and female layers made of PTFE and PEEK.

can be made from layers of elastomers such as Viton, FKM, Neoprene, EPDM, etc. The bottom and top layers have one flat surface on both ends, rather than the V shape, and other V-shaped layers are located in between. The combination of construction materials defines the resistance of the V-pack against pressure, temperature, and fluid. A V-pack is considered a very high technology sealing that is not applicable to valves in refineries, petrochemical plants, or for use in the topside offshore sector of the oil and gas industry. A V-pack is a common packing for valves installed and used in the subsea sector of the oil and gas industry.

3.6.5 Compression packing

One of the oldest methods used to seal components is compression packing. Compression packing is made of different rings that are inserted in an annular space called a stuffing box. Sealing is achieved by tightening the gland flange bolts and the gland that provides loads on the packing rings. The sealing capability of the compression packing is defined as the ability to expand laterally against the stem and stuffing box wall when stressed by tightening the gland. The stress applied laterally by the compression packing stem sealing as a result of the axial stress applied from the gland and gland bolting depends on the Poisson's ratio of the material (μ) which is expressed in Formula 3.1.

Lateral stress calculation in packing

$$\delta_L = \delta_a \left(\frac{1 - \mu}{\mu} \right) \tag{3.1}$$

where:

δ_L: Lateral stress;
δ_a: Axial stress;
μ: Poisson's ratio.

As an example, a soft rubber material with a Poisson's ratio of 0.5 can create axial stress equal to lateral stress. Graphite is less flexible than rubber material and has less lateral expansion compared to elastomeric rubber under a specific axial force from the gland bolts. Fig. 3.16 illustrates compact packing rings used for stem sealing.

FIGURE 3.16 Compact packing ring. *Courtesy: Shutterstock.*

The other important consideration is the effect of stem packing rings on the packing operating torque. Valve torque is defined as the measurement of the force required to operate the valve between open and closed positions. Different packing parameters affect valve operation torque, such as the packing material, number of packing rings, amount of load applied by the gland on the packing rings, stem surface finish, temperature, etc. Increasing the number of packing rings can help reduce fugitive emission, although it increases the torque required for valve operation. Formula 3.2 is used to measure the required force to overcome the packing friction.

Calculation of required force to overcome packing friction

$$F = \pi \times d \times H \times GS \times \mu \times Y \tag{3.2}$$

where:

F: Force required to overcome total packing rings friction (N);
d: Stem diameter (m);
H: Total packing rings height (m);
GS = Compressive stress on the packing (axial gland stress on the packing) (N/m^2);
μ = Packing coefficient of friction (dimensionless);
Y = Ratio of axial load transference (e.g., 0.5).

The total packing forces on the stem (parameter N) due to all the packing rings can be calculated through Formula 3.3.

$$N = \pi \times d \times H \times \mathrm{GS} \times Y \tag{3.3}$$

where:

N: Total packing forces on the stem (N);
d: Stem diameter (m);
H: Total packing rings height (m);
GS: Compressive stress on the packing (axial gland stress on the packing) (N/m^2);
Y: Ratio of axial load transference (e.g., 0.5).

Fig. 3.17 illustrates seven packing rings around a valve stem that is pressurized by fluid on one side and by gland pressure from the other side. The total of the lateral forces applied by the packing rings on the stem is illustrated in red in Fig. 3.17. The amount of load applied from the packing rings laterally on the valve stem is not distributed equally. The amount of lateral force applied to the stem is the highest on the two packing rings installed under the gland flange. The figure shows that 70% of the lateral sealing force

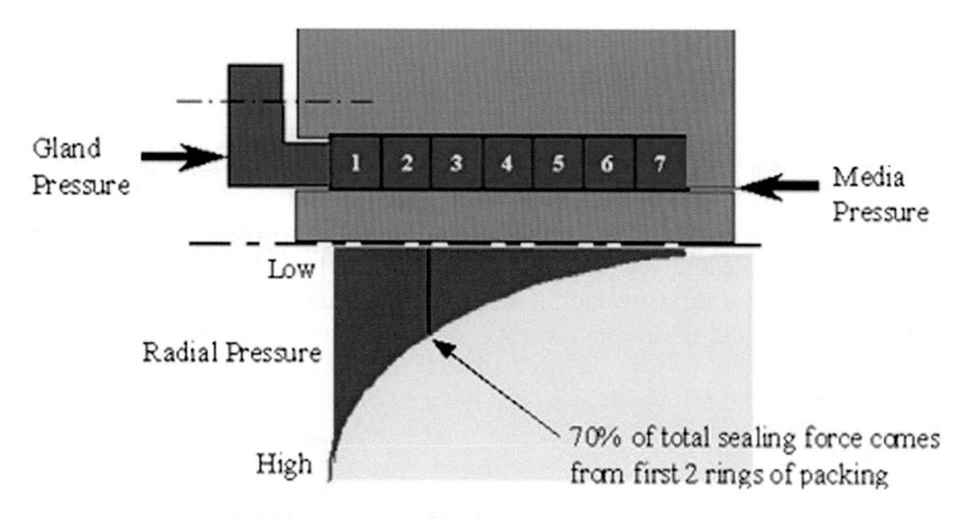

FIGURE 3.17 Stem packing load distribution on the stem.

from the seven packing rings is coming from just two of the rings installed below the gland flange.

Question: The stem diameter of a gate valve is 11.1 mm. The stem has three packing rings and the height of each ring is 20 cm. The amount of compression stress on the packing from the gland, gland flange, and bolting is 2000 psi. How much lateral force is applied from these three packing rings to the stem of the valve? How much load is applied from the packing located just under the gland, assuming that 50% of the packing loads on the stem are coming from the packing just under the gland?

Answer:

d = stem diameter = 11.1 mm = 0.111 m, H = total packing height = 3 × 20 cm = 0.6m, GS = compression stress on the packing = 2000 psi = $1.39 \times 10^7 \frac{N}{m^2}$, 1 psi is equal to 6894.76 $\frac{N}{m^2}$

$N = \pi \times d \times H \times GS \times Y = 3.14 \times 0.111\ m \times 0.6\ m \times 1.39 \times 10^7 \frac{N}{m^2} \times 0.5 = 0.14 \times 10^7\ N.$

The total amount of load from the three layers of packing to the stem is equal to 0.14×10^7 N. If just 50% of this load is coming from under the gland flange, then the amount of load from the packing under the gland flange to the stem would be 0.07×10^7 N.

The axial gland stress (GS) on the packing is calculated through Formula 3.4.

$$GS = \frac{\frac{T.n}{0.2.b}}{\frac{\pi(D^2 - d^2)}{4}} \tag{3.4}$$

where:

T = The torque applied to the gland flange bolts;
n = The number of bolts;
b = Nominal bolt diameter;
D = Packing box diameter;
d = Stem diameter.

Compression packing can be energized by live loads, such as the spring loads on the gland flange (see Fig. 3.18). Usage of spring on the gland flange to provide constant and extra load on the packing is considered one of the simplest forms of live load application on the gland flange. Live load spring provides more and sufficient load on the gland flange and the packing, even if the gland flange load is not sufficient after some time of operation. In fact, live load spring compensates for lack of gland flange load on the packing due to a variety of factors, such as misalignment of the gland flange.

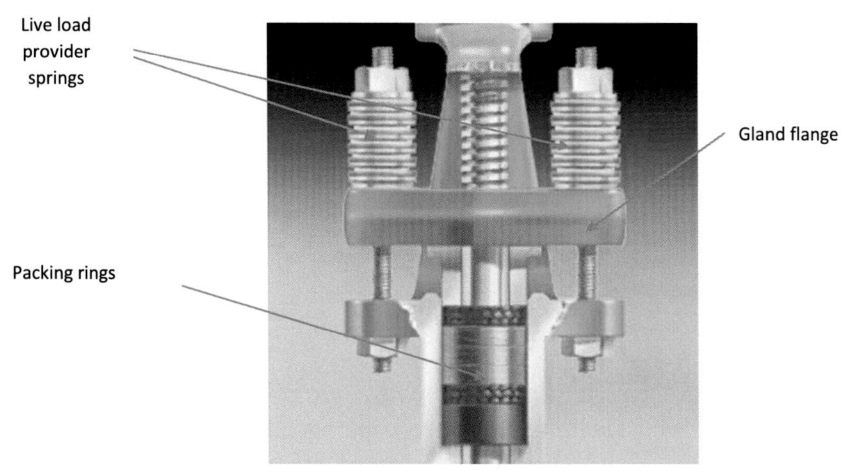

FIGURE 3.18 Live load gland flange.

3.6.6 Bellows

Bellows are metallic seals that are used for stem seals when very strong and strict sealing performance is required. In fact, bellows seals could be a good choice of valve stem seal in toxic and flammable liquids as well as high-pressure fluid. The bellows is welded on both the valve stem and bonnet and provides zero fugitive emission. The bellows cartridge has numbers of convolutions that are expanded or compressed during stem movement. The constant expansion and compression of the bellows creates fatigue loads that could cause bellows failure during operation. Using multiple bellows and reducing the length of the stem movement and valve stroke can save the bellows life against fatigue stress. Noticeably, the total flexibility of the bellows depends on its number of convolutions. Fig. 3.19 illustrates bellows stem seals for a control valve.

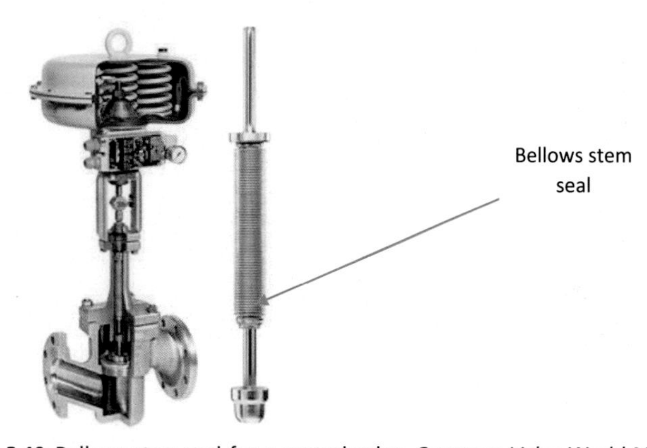

FIGURE 3.19 Bellows stem seal for a control valve. *Courtesy: Valve World Magazine.*

3.7 Valve sealing materials

This section focuses on the materials that can be used for different types of valve seals, and the advantage and disadvantage of each one. In general, soft or metallic materials, or a combination of soft and metallic materials, are used for valve sealing. The choice of sealing materials depends on different parameters, such as the type of valve, the operating and design pressure and temperature, the application of the valve (subsea or topside), the fluid service of the valve, etc.

3.7.1 Stem sealing materials

3.7.1.1 Graphite

Gate and globe valve are typically equipped with nonmetallic graphite packing (see Fig. 3.20, left side) or PTFE (Teflon). Graphite packing has different advantages; it is highly resistant to corrosive and chemical fluids and fire resistant and can be used in extreme high- and low-temperature conditions. Due to these advantages, graphite is typically the preferred packing or stem sealing for gate and globe as well as butterfly valves. Graphite packing can be used for extreme temperatures that could be as low as −250°C and as high as 500°C. Graphite can be used in two forms: dye-formed and braided. However, graphite packing has two disadvantages; first, graphite can provide high friction with the valve stem compared to other nonmetallic or soft sealing materials. The second disadvantage is that graphite is a very Nobel material that could cause galvanic corrosion of the valve stem in the presence of certain electrolyte fluids such as water. The high purity of the graphite (more than 98% or 99% pure) is a key consideration for graphite packing design and selection. Graphite may be required to have less than 50 ppm chloride and less than 700 ppm sulfur. The reason for putting a limitation on the chloride content is that it can form acid and lead to corrosion of the packing.

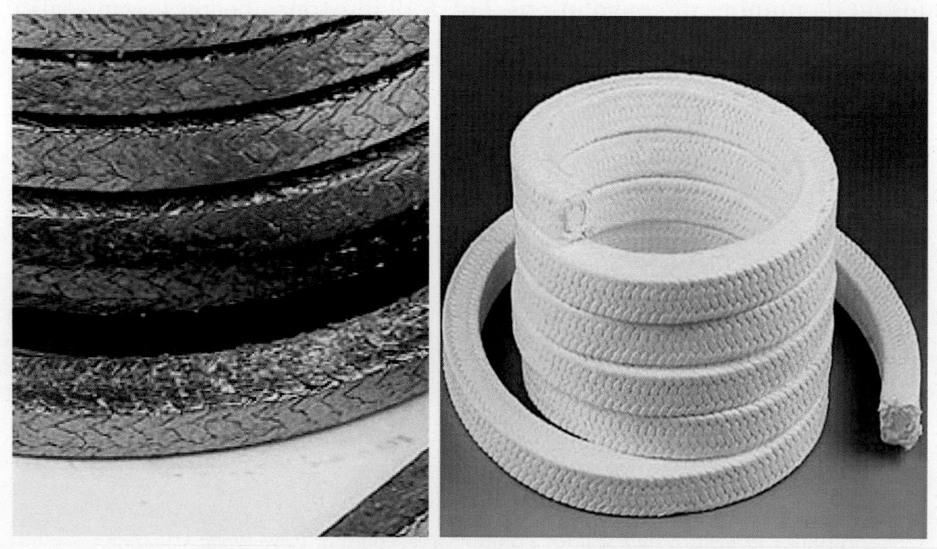

FIGURE 3.20 Graphite packing (left) and PTFE packing (right).

Sulfide should be limited since it can oxidize and make SO_2, which disturbs the tightness of the packing and causes corrosion.

3.7.1.2 PTFE (Teflon)

PTFE is the shortened name of the chemical polytetrafluoroethylene; Teflon is its tradename. PTFE is a type of thermoplastic material that becomes moldable at a certain elevated temperature. PTFE may be preferred over graphite due to the two disadvantages of graphite mentioned above, including high friction with the stem and galvanic corrosion of the stem. PTFE has high corrosion resistance due to its chemical properties and self-lubricating quality, and it provides very low stem friction unlike graphite. However, PTFE is not fire resistant and has a more limited temperature range. PTFE packing, as illustrated in Fig. 3.20 (right side), is white in color. PTFE can provide tighter sealing and is less likely to leak compared to graphite packing. It should be noted that PTFE can be reinforced with graphite or glass fiber, which makes it more resistant to higher pressure and temperature. Reinforced PTFE is abbreviated as RPTFE.

3.7.1.3 Elastomeric seals

Elastomeric materials such as Viton are another type of stem seal material that may be used sometimes for the ball valves. In certain circumstances, a ball valve may contain triple stem seals in the shape of an O-ring. Elastomeric seals can provide leakage class/ tightness at the same level as PTFE stem sealing or packing. Figs. 3.21 and 3.22 illustrate

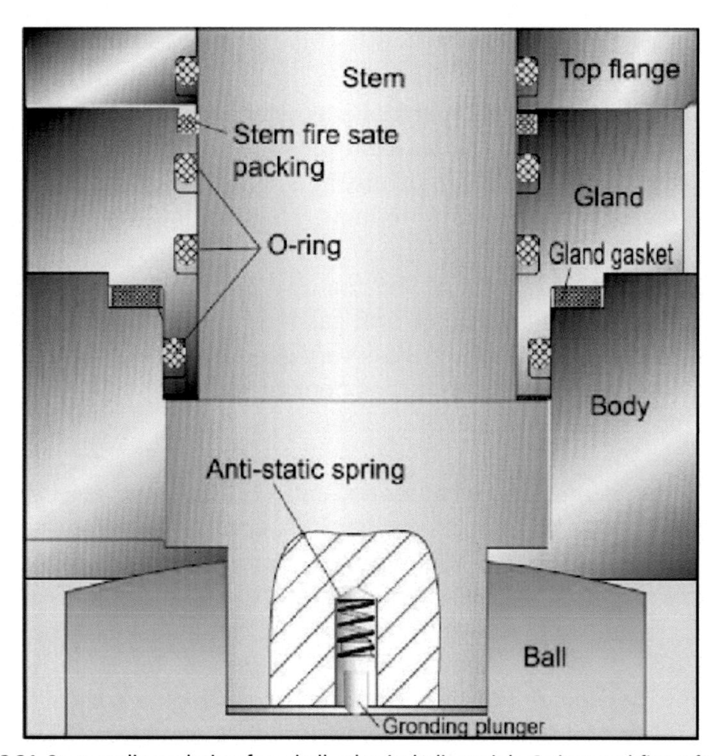

FIGURE 3.21 Stem sealing solution for a ball valve including triple O-rings and fire-safe packing.

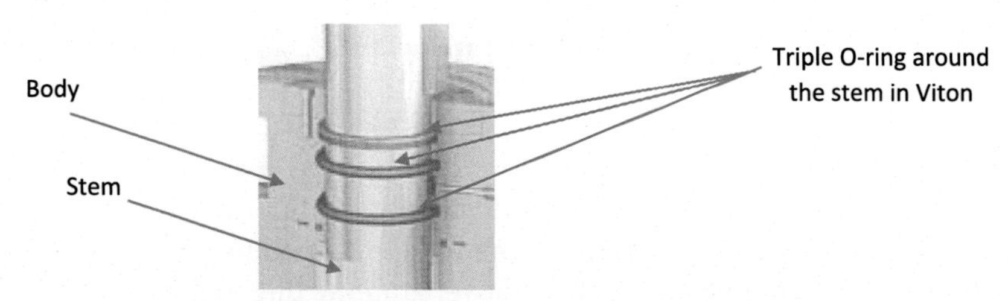

FIGURE 3.22 Stem sealing for a ball valve including triple O-ring.

a ball valve stem sealing solution that includes triple O-rings around the stem as well as graphite stem packing to provide stem sealing in case of fire, as the triple O-rings would melt.

Different O-ring materials can be used for industrial valves, which are summarized as follows:

1. **Buna N/Nitrile rubber/NBR:** This type of synthetic rubber is very common for valve sealing and is used for both stem sealing and body and bonnet sealing in the oil and gas industry. It can be used in an operating temperature range between −20°C and 90°C. This material has very good resistance to chemicals and hydrocarbons, such as oil, and very good abrasion resistance. However, the resistance of Buna N to ozone, weather, and UV rays is poor.
2. **HNBR:** Hydrogenated Buna N (HNBR) has higher temperature resistance and ozone resistance compared to Buna N. HNBR can be used in a temperature range between −40°C and 150°C.
3. **EPDM:** This is a synthetic rubber made from the copolymerization of ethylene and propylene. The full name of this material is ethylene propylene diene monomer. EPDM can be used in a variety of temperatures ranging from −50°C to 150°C. It has excellent resistance to weather, ozone, and heat. However, it has poor resistance to oil and gas and petroleum products. That could be a reason why this type of rubber is not popular for valves in the oil and gas industry.
4. **Neoprene:** This type of synthetic rubber material is not common for valves and actuators in the subsea sector of the oil and gas industry. It is produced by the polymerization of chloroprene. Neoprene has very good abrasion, water, ozone, UV rays, and heat resistance. However, it has poor resistance to acids and chloride.
5. **Silicon:** This type of rubber material is excellent in high-temperature conditions and may be used in a wide range of temperatures from −60°C to 220°C. It provides very good UV rays, ozone, and weather resistance. It has poor resistance to oil, so it is not common in the oil and gas industry, especially in the subsea sector.
6. **Viton:** Viton is the trademark of the fluoroelastomers, also known as FKM. This material is very common for the sealing of industrial valves (mainly stem sealing) in the oil and gas industry, especially subsea, and can be used in the temperature range of −40°C to 180°C. Viton provides outstanding resistance to the weather, UV rays, and ozone. In general, FKM does not provide as high a corrosion resistance as HNBR to chemicals and hydrocarbon. But Viton does provide very good pressure,

temperature, and corrosion resistance to hydrocarbon fluid, and it is relatively inexpensive.

3.7.1.4 Lip seal materials

Lip seals can be used for sealing around the valve stem in applications where elastomeric O-ring materials are not robust enough. As an example, O-rings may fail in high-pressure gas applications due to lack of AED property, so lip seals can be used alternatively. Lip seals are made of a soft material and a metallic spring material. The soft material could be elastomeric, such as Viton, Buna N, etc., or thermoplastic, such as PTFE. The metallic ring could be in stainless steel 316 for refineries and petrochemical plants. However, a more corrosion-resistant spring material, such as a cobalt alloy like Elgiloy (UNS R30003) or a nickel alloy such as Inconel 625 or Hastelloy C276, is required for lip seals in the offshore sector of the oil and gas industry.

3.7.1.5 Bellows materials

Bellows are in metallic materials. Stainless steel 316 bellows material is more common for refineries and petrochemical plants where less corrosion-resistant alloys are used. Conversely, some nickel alloys such as Inconel 625, Hastelloy C or C276 could be more applicable for offshore. A bellows stem seal (see Fig. 3.23) is used for large subsea ball valves especially those installed at deepwater depths such as 1 km.

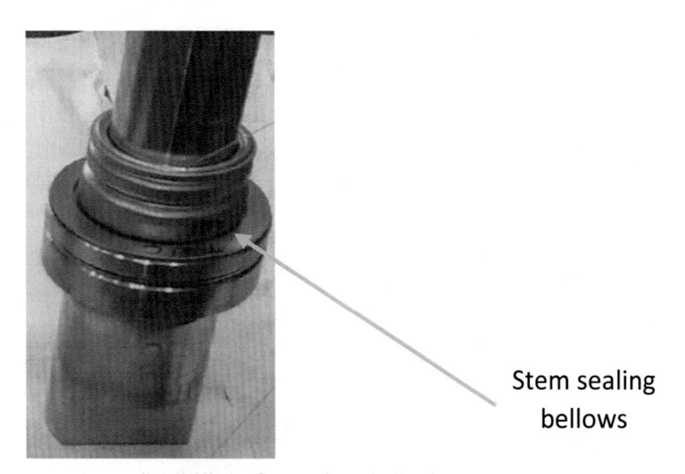

FIGURE 3.23 Stem sealing bellows for a subsea ball valve. *Courtesy: Galperti Engineering.*

3.7.1.6 V-packs and Glyd rings (subsea valve seals)

Subsea valves in general have a more complicated stem sealing system compared to topside valves. Subsea valve seals typically combine two or three different sealing mechanisms and provide sealing not only against the leakage of internal fluid to the sea, but also against the leakage of seawater to the valve internals. A combination of

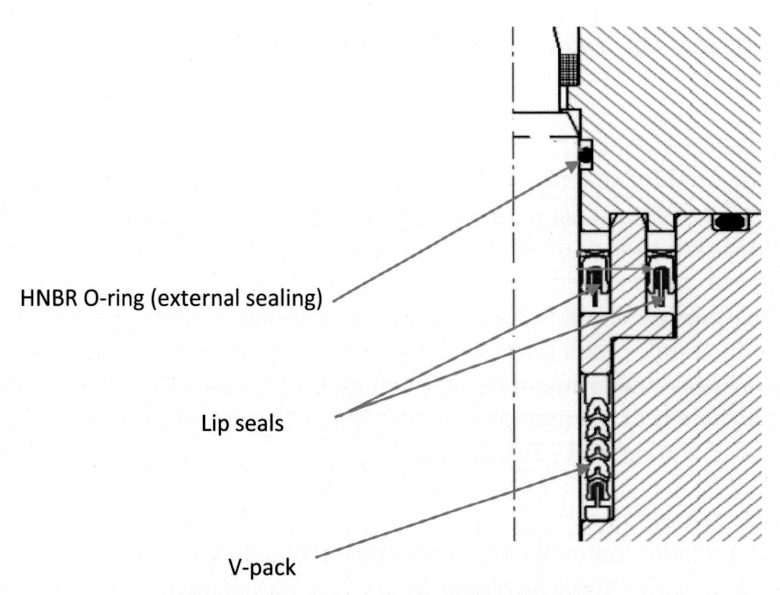

FIGURE 3.24 Stem sealing for a slab gate valve.

lip seals, V-packs, O-rings, and bellows are used for subsea valve stem seals. As an example, Fig. 3.24 illustrates stem seals for a slab gate valve that include a V-pack as the primary fluid barrier, a lip seal as a secondary fluid barrier, and an O-ring as an external seawater barrier. V-packs are typically made of layers of soft materials such as thermoplastic seals (PTFE, PEEK) or elastomeric seals such as Viton.

A Glyd ring (see Fig. 3.25) is the trademark name of a double-acting O-ring with an energized piston seal used for dynamic applications such as stem sealing. A Glyd ring is proposed in some slab gate valves as the primary barrier against seawater ingress to the valve stem areas instead of an O-ring. This sealing has many advantages, such as low friction, minimum breakout force, and high wear resistance. Glyd rings can also be used for actuator cylinder sealing.

FIGURE 3.25 Glyd ring.

3.7.2 Body/bonnet sealing materials

3.7.2.1 Gaskets

Gaskets are widely used for sealing the body and bonnet or body piece components in both topside and subsea valves. Graphite or other nonmetallic gaskets are common for low-pressure class valves in utility services such as water in refineries and petrochemical plants. An example of a soft material gasket could be a compressed non-asbestos sheet gasket material produced from aramid and/or inorganic fibers, and a high-quality filler bonded with acrylonitrile-butadiene rubber (nitrile). Spiral wound gaskets could be used for valves in low-pressure classes such as ASME pressure class 150 or 300. Spiral wound gaskets for valves used in refineries and petrochemical plants could have a stainless steel 304 or 316 winding and inner ring with a carbon steel outer ring. However, more exotic materials for spiral wound gaskets can be used in the offshore industry, such as Inconel 625 metallic winding and rings as an example. RTJ gaskets used for high-pressure piping are typically made of austenitic stainless steel grades such as 304 and 316 for valves installed in refineries and petrochemical plants. More exotic RTJ gasket materials, such as Inconel 625, are common in the offshore sector of the oil and gas industry, especially subsea.

3.7.2.2 O-rings

O-rings in elastomeric materials may be used between the body and bonnet pieces of some valves, such as butterfly valves, in utility (nonprocess) services. It should be noted that O-rings and lip seals are used for sealing the internal parts of the valves, which are not discussed in this chapter since those seals do not affect fugitive emission.

3.8 Questions and answers

1. Leakage of gas from which valve seal type is not counted as fugitive emission?

A. Stem sealing

B. Body/bonnet and body pieces gaskets

C. Seat insert

D. Plug Loctite

Answer: Leakage of gas from the stem sealing typically enters the environment and is considered a fugitive emission. Leakage of gas from the body/bonnet and body pieces gaskets, like leakage from the stem, is released to the environment and is counted as a fugitive emission. Leakage from the seat insert flows to the body cavity of the valve and not to the environment. Thus, leakage from the seat insert is not counted as a fugitive emission. Plugs are installed on the valve cavity for drainage of the overpressurized fluid from the cavity. Loctite is a type of sealant used on the plug threads to prevent leakage of the fluid from the plugs. Leakage through the Loctite and plug enters the environment and is counted as a fugitive emission.

2. An industrial ball valve is placed in oil service with an operating temperature of 50°C and a design pressure of 30 bar. Which type of stem seal and sealing between the body pieces are the most economical and technically accepted solution?
   - **A.** Ring-type joint (RTJ) metallic gasket and lip seal stem seal
   - **B.** Spiral wound gasket and O-ring stem seal
   - **C.** Flat gasket and O-ring stem seal
   - **D.** Spiral wound gasket and lip seal stem seal

   Answer: The design pressure is 30 bar, so according to ASME B16.34, the valve design standard, the pressure class is 300. Ring-type joint gaskets are typically selected for high-pressure classes such as CL600 and above. Thus, option A is not correct. Flat gaskets are typically used for low-pressure class services such as CL150 and utility services such as water. Thus, a flat gasket is under design for the given application. Under design means poor design below the required application. The correct type of gasket in this case is a spiral wound gasket, which is suitable for CL300 and oil service. Now, the answer should be selected from one of the options B and D. Both lip seals and O-rings can be selected for stem sealing. However, lip seals are more expensive and could be over design. Over design means to design in a manner that is excessively complex or exceeds the required application. Lip seals are proposed for antiexplosive decompression (AED) applications where the fluid service in the valve is gas, and in high-pressure classes above CL300. Rapid reduction of the gas service pressure leads to ingress of gas into the elastomer seal which expands it and finally causes sealing failure. In this case, the fluid is oil, so there is no requirement to have an AED sealing. Thus, an O-ring stem seal is sufficient. If AED is a requirement, it is possible to select an AED O-ring. Thus, option B is correct.

3. Which sentence is not correct regarding a lip seal?
   - **A.** A lip seal could be selected for the stem as well as the body and seat sealing of industrial valves in all sectors of the oil and gas industry.
   - **B.** Lip seals provide sealing just in one direction.
   - **C.** An austenitic stainless steel spring in 316 for a lip seal used for a subsea valve is a suitable material choice.
   - **D.** A lip seal is suitable for a gas service with an operating temperature between −29°C and 180°C.

   Answer: Option A is correct, since a lip seal can be used for stem sealing as well as body and seat sealing of valves in different sectors of the oil and gas industry, such as refineries, petrochemical plants, and offshore facilities. Option B is also correct, since a lip seal cannot provide sealing from both directions. Option C is not correct, since austenitic stainless steel 316 is susceptible to chloride and prone to pitting and stress cracking corrosion in seawater. It should be noted that seawater is a chloride-containing environment. Option D is correct, since lip seals are suitable for operating temperatures between −29°C and 180°C.

4. Which sentence is not correct about subsea valve sealing?
   A. Spiral wound gaskets are not used for subsea valves.
   B. A V-pack can be used for subsea valve stem sealing, and is comprised of thermoplastic materials.
   C. Bellows are not recommended for linear gate valves due to their high friction with the valve stem.
   D. A lip seal is only applicable for valve stem sealing.

Answer: Spiral wound gaskets are made of metallic rings and graphite fillers according to ASME B16.20 (see Fig. 3.26). This type of gasket is suitable for low-pressure classes, such as class 150 equal to a nominal pressure of 20 bar and class 300 equal to a nominal pressure of 50 bar. Subsea valves based on API 6A/17D have much higher-pressure classes, so spiral wound gaskets are not applicable. Therefore, option A is correct. Option B is also correct, since V-packs are used for the stem sealings of subsea valves, which could contain male and female layers of PEEK and PTFE as thermoplastics. It is essential to remember that a V-pack can be used for sealing the seat and body of gate and ball valves. Therefore, the usage of V-packs is not limited to stem sealing in subsea valves. Option C is correct, since a bellows seal can create high friction with the stem, especially in a gate valve in which the valve stem moves linearly and creates a high amount of friction with metallic sealing such as a bellows. Option D is wrong, since lip seal usage is not limited to stem sealing, and a lip seal, like a V-pack, can be used for sealing between the seat and body of ball and gate valves.

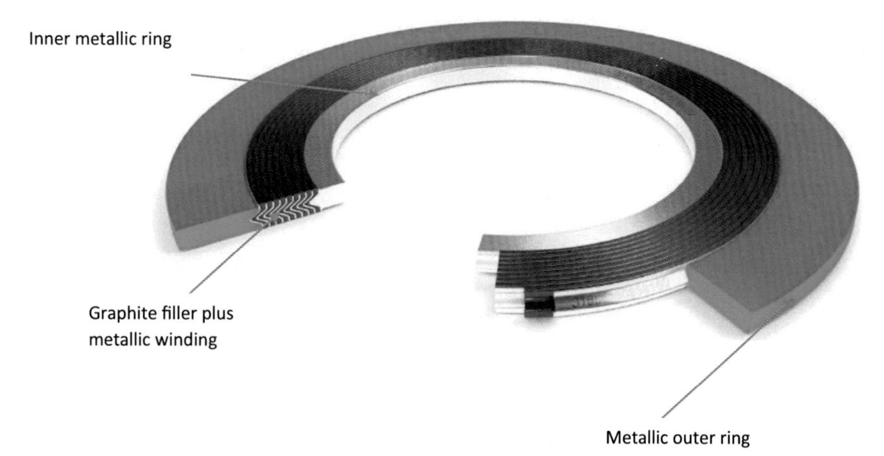

FIGURE 3.26 Spiral wound gasket.

5. Which parameters do not affect sealing selection for industrial valves?
   A. Type of motion of the components in contact with the sealing
   B. Operating pressure and temperature of the valve
   C. Requirements for AED and cost
   D. Velocity of the fluid in the valve

Answer: The type of motion of the components in contact with the sealing affects sealing selection. Static seals located between two static components are less prone to wearing compared to dynamic seals located between at least one moving component. Therefore, option A is correct. Operating pressure and temperature affect valve sealing material selection. As an example, PTFE has limited pressure and temperature resistance compared to a lip seal. Thus, option B is correct as well. Option C is also correct, since economic factors and AED requirements affect sealing selection. AED requirement may lead to the selection of lip seals instead of O-rings. Economic factors should be considered for valve sealing selection but the cost of sealing is probably not that important since it is minor compared to the total cost of the valve, which is mainly in steel material. Option D is not correct, since the velocity of the fluid inside the valve does not affect valve sealing selection. Fig. 3.27 illustrates some of the parameters that affect valve sealing selection.

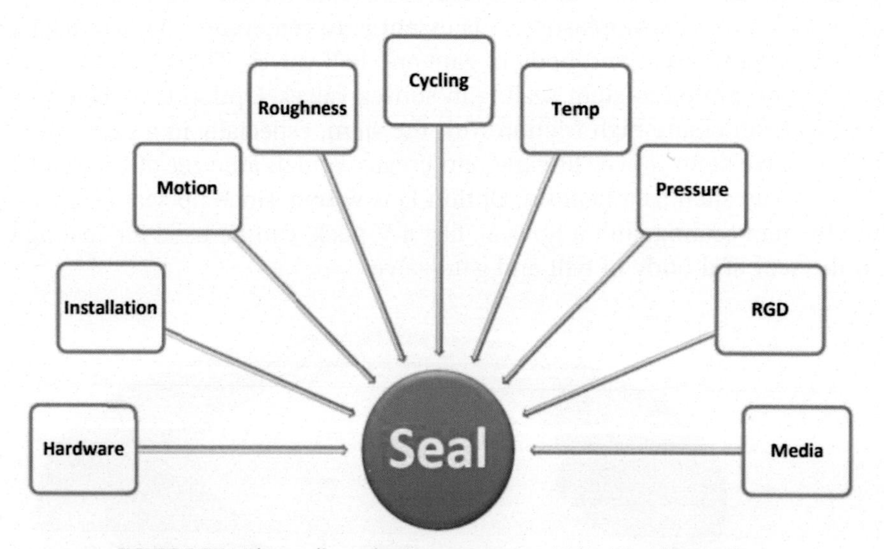

FIGURE 3.27 Valve sealing selection parameters. *Valve World Magazine.*

Bibliography

[1] American Petroleum Institute (API) 17D. Design and operation of subsea production systems, subsea wellhead and tree equipment. 2nd ed. Washington, DC, USA: API; 2011.

[2] American Petroleum Institute (API) 6A. Specification for wellhead and tree equipment. 21st ed. Washington, DC, USA: API; 2018.

[3] American Petroleum Institute (API) 598. Valve inspection and testing. 8th ed. Washington, DC, USA: API; 2004.

[4] American Society of Mechanical Engineers (ASME). Metallic gaskets for pipe flanges. ASME B16.20. New York, NY: ASME; 2017.

[5] American Society of Mechanical Engineers (ASME). Non-metallic flat gaskets for pipe flanges. ASME B16.21. New York, NY: ASME; 2016.

[6] Fluid sealing association. Compression packing technical manual. 4th ed. 2018 [PA. USA].

[7] International Organization of Standardization (ISO) 5208. Industrial valve- pressure testing of metallic valves. 4th ed. 2015 [Geneva, Switzerland].

[8] Mokhatab S, Poe WA. Handbook of natural gas transmission and processing. 2nd ed. USA: Elsevier; 2012. p. 619–78. https://doi.org/10.1016/C2010-0-66115-3.

[9] Nesbitt B. Handbook of valves and actuators: valves manual international. 1st ed. Oxford, UK: Elsevier; 2007.

[10] Smit P, Zappe RW. Valve selection handbook. 5th ed. New York, NY: Elsevier; 2004.

[11] Sotoodeh K. Soft materials selection, application and the implications on valve design. Valve World Mag 2016;21(8):51–5.

[12] Sotoodeh K. Why are Butterfly valves a good alternative to ball valves for utility services in the offshore industry? Am J Ind Eng 2018;5(1):36–40. https://doi.org/10.12691/ajie-5-1-6.

[13] Sotoodeh K. A review of valve stem sealing to prevent leakage from the valve and its effect on valve operation. J Fail Anal Prev 2020. https://doi.org/10.1007/s11668-020-01050-1. Springer.

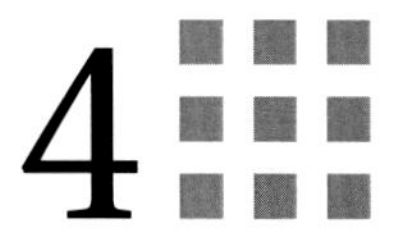

4 Valve fugitive emission test standards, specifications, and laws

4.1 Introduction and history

As discussed in Chapter 2, the highest percentage of leakage in industrial plants is generated from industrial valves. Different regulatory bodies and governmental organizations are concerned about fugitive emission. Valve and packing manufacturers are looking for approaches to manufacture and supply packings and valves with Low-E (emission) requirements to comply with rules, regulations, standards, and end users' specifications. There are different fugitive emission standards, laws, and customer specifications for valves, each with different parameters. The American Petroleum Institute (API) and International Organization for Standardization (ISO) are considered the primary fugitive emission standards. Fig. 4.1 summarizes the most important ISO and API international standards for valve fugitive emission tests.

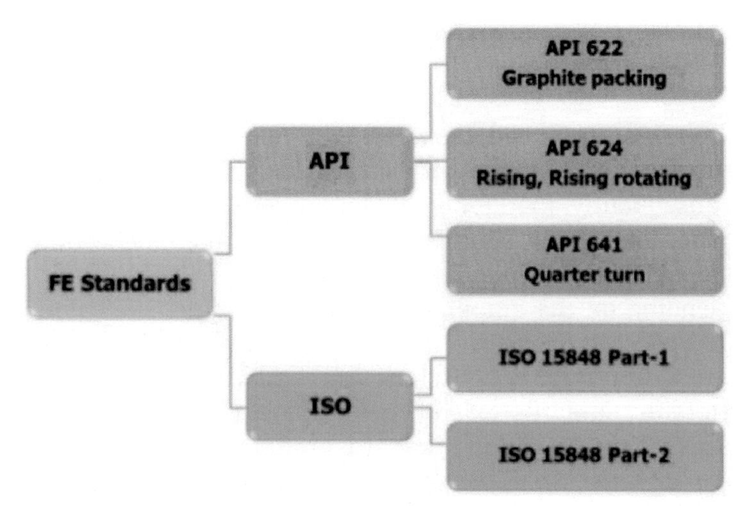

FIGURE 4.1 The most important API and ISO standards for valve fugitive emission tests.

Two main end-user specifications are Shell oil company's material and equipment standards and codes (MESC) SPEC 77/312 and Chevron's fugitive emission specifications. Two essential federal fugitive emission rules and regulations are TA Luft (VDI

Prevention of Valve Fugitive Emissions in the Oil and Gas Industry. https://doi.org/10.1016/B978-0-323-91862-6.00003-4

2440) and the United States Environmental Protection Agency (EPA) 40, parts 60/63 (EPA method 21). Fig. 4.2 summarizes the most important fugitive emission laws, standards, and customer specifications for valves in the oil and gas and chemical industries.

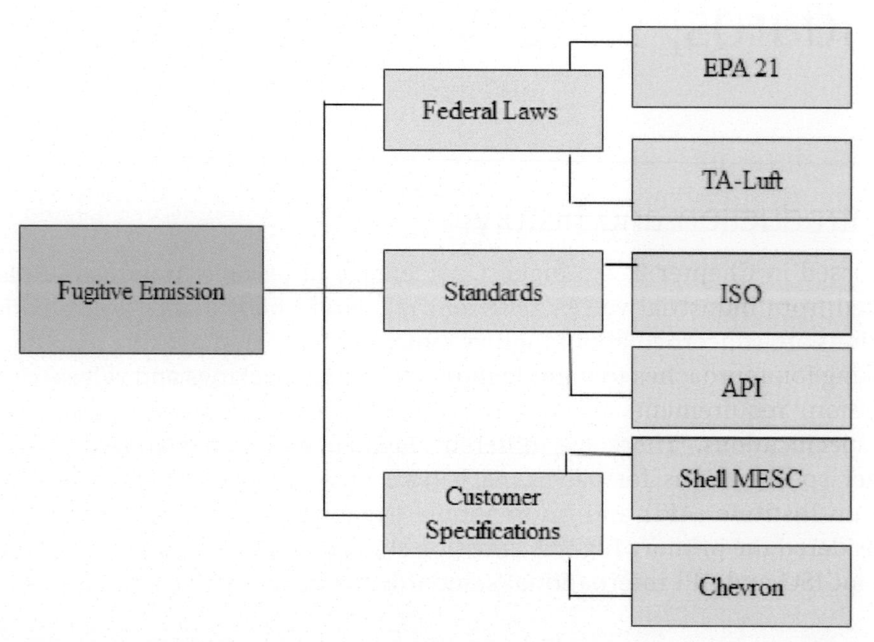

FIGURE 4.2 Major federal laws, standards, and customer specifications related to valve fugitive emissions in the chemical and oil and gas industries.

The United States introduced the Clean Air Act in 1990 with the objective of reducing the emission of volatile organic compounds (VOCs) and chemicals into the atmosphere. Consequently, the EPA created a process for monitoring emissions and scheduling repairs for leaking components and equipment. The frequency of leak detection and repair (LDAR) depends on the number of valves that are leaking above the limits. Depending on the amount of fugitive emission, the frequency of LDAR could vary from a month to 1 year.

The importance of reducing emissions from valves is one of the main reasons why many different standards exist to test and evaluate fugitive emission from valves. All of the abovementioned standards and specifications were written and created based on different areas of expertise, considerations, and perspectives. It is a very challenging task to compare these standards all together, since they all have different test procedures. But the most important differences between all of these standards can be summarized in terms of five main variables: test fluid or medium, detection methods for leakage, limitations of the leakage, type of valve stroke, and number of cycles.

4.2 Standards and regulations review

4.2.1 API standards

API has published two standards addressing fugitive emissions from valves and one standard from the valve stem packings. The main purpose of these standards is to establish a uniform procedure to evaluate the sealing ability of the valve stem packing for process applications where fugitive emission is a concern. The testing programs that are introduced in these standards are a basis for the life cycle performance of the valve stem packing in terms of emission concerns. These standards force the valve manufacturers to implement the standards and test the seals and packing designs to make sure that the valves can achieve the specified fugitive emission levels.

4.2.1.1 API 622 scope

The API 622 standard provides a comprehensive procedure for conducting a fugitive emission test on stem packing. The fugitive emission test includes test arrangement, leak test equipment selection and calibration, packing selection, and installation. The important point is that this standard is not limited to a fugitive emission test. API 622 also establishes the requirements and parameters for the corrosion test as well as packing material composition and properties. Additionally, this standard covers all of the stem motion types, such as rising stem, rotating stem, and rising–rotating stem. In fact, this standard is designed to provide a clear road map for packing qualifications and to provide confidence to end users about the low fugitive emission valves that are tested based on the standard. It is important to note that this is a strict standard for testing the packing, but not the valves.

4.2.1.1.1 API 622 fugitive emission test

4.2.1.1.1.1 Test fixture The test arrangement and key components are shown in Fig. 4.3, extracted from API 622. Fig. 4.4 provides a simplified, general arrangement for the fugitive emission test fixture according to the API 622 standard. The test fixture should be equipped with an actuator on the valve shaft equivalent on the right side of Figs. 4.3 and 4.4 to stroke the test stem for a mechanical cycle. The stroke distance or length, degree, and speed are defined in the standard. The key point is that the actuator should not apply any side load to the valve stem. Two thermocouples are connected to the test fixture to monitor and measure the temperature; one is connected to the test chamber, which contains pressurized methane, and the other is connected to the stuffing box area. The fixture is heated by using an external heat source in the form of a blanket or heating coils.

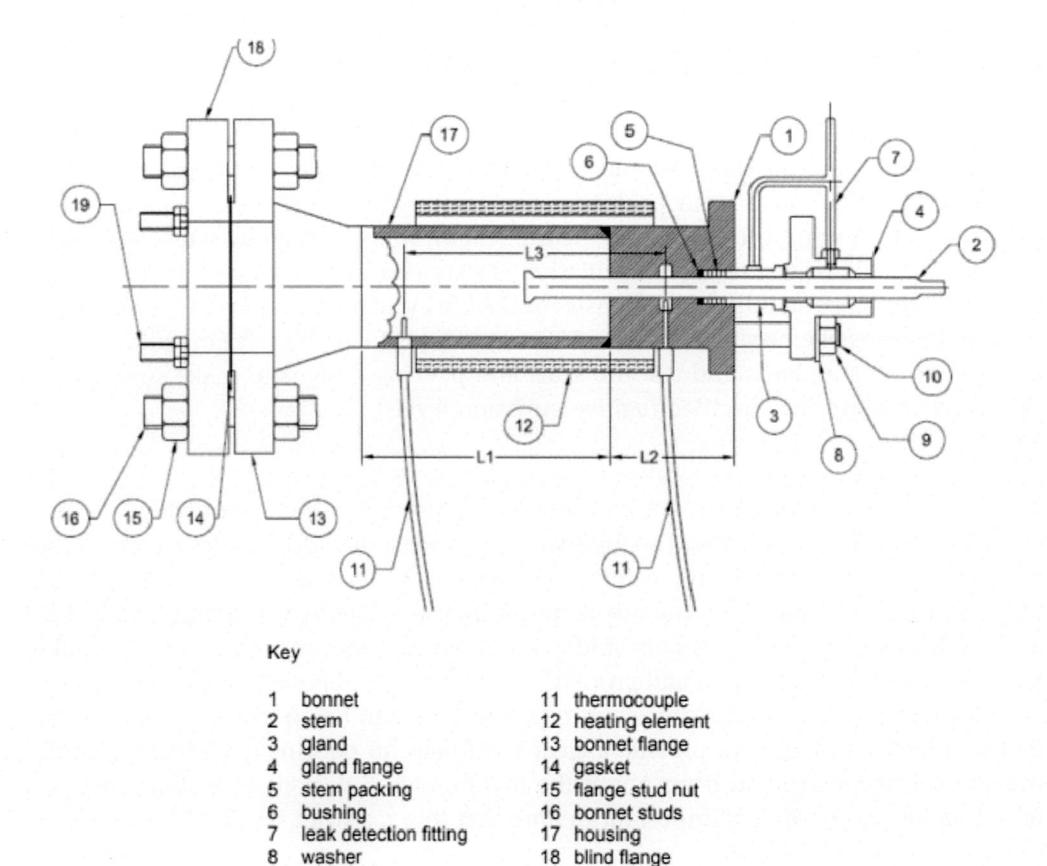

FIGURE 4.3 Fugitive emission test arrangement as per API 622.

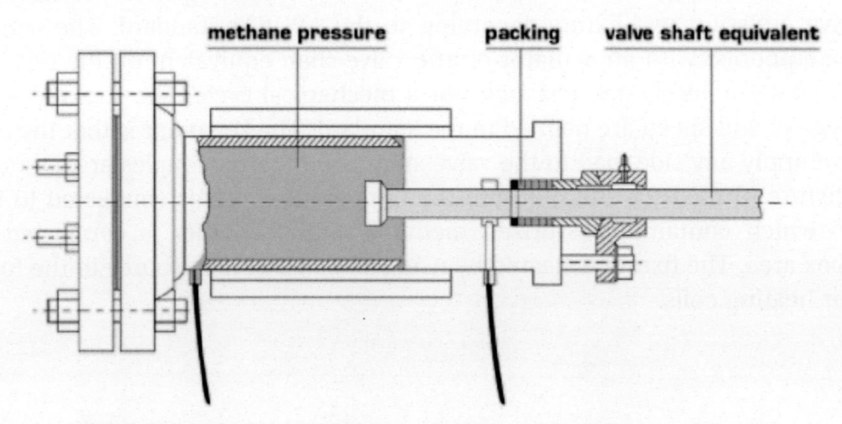

FIGURE 4.4 General arrangement for a fugitive emission test as per API 622.

Different dimensions for the test parameters given in Fig. 4.3, such as L_1, L_2, and L_3, are specified in the test standard. The dimensions of the test fixture or arrangement are given for two types of packing cross section: 1/4″ and 1/8″. The formulae for calculating the packing size, also called the packing cross section area, as well as the number of packing rings are provided in Chapter 1. Table 4.1, extracted from API 622, provides L_1, L_2, and L_3 dimensions for both 1/4″ and 1/8″ packing dimensions. The table also proposes the material for the gland nuts. The gland bolts should be compatible with the gland nuts and in ASTM A193 grade B7. B7 bolt material is categorized as a low alloy steel containing chromium and molybdenum alloys. Other materials in the test fixture are as follows: The gland material should be in forged carbon steel in ASTM A105 or cast material in ASTM A216 WCB. The stem should be in 13 chromium stainless steel ASTM A182 F6a with a roughness of 15–28.8 Rc equal to 200–275 HB. The body of the test fixture is made of carbon steel seamless pipe having schedule 80 as minimum.

Table 4.1 Test fixture (arrangement) dimensions as per API 622.

Item	1/8 in. Packing fixture	1/4 in. Packing fixture
L1	127.0 mm (5.00 in.)	193.5 mm (7.62 in.)
L2	63.5 mm (2.50 in.)	139.7 mm (5.50 in.)
L3	127.0 mm (5.00 in.)	260.3 mm (10.25 in.)
Gland nuts	ASTM A194 grade 2H	ASTM A194 grade 2H

The test defines the test arrangement and fixture dimensions very strictly. The standard provides very strict dimensions for the bonnet as per Fig. 4.5 and Table 4.2. In addition, the stem/stuffing box dimensions as per Table 4.3 are clearly defined in this standard.

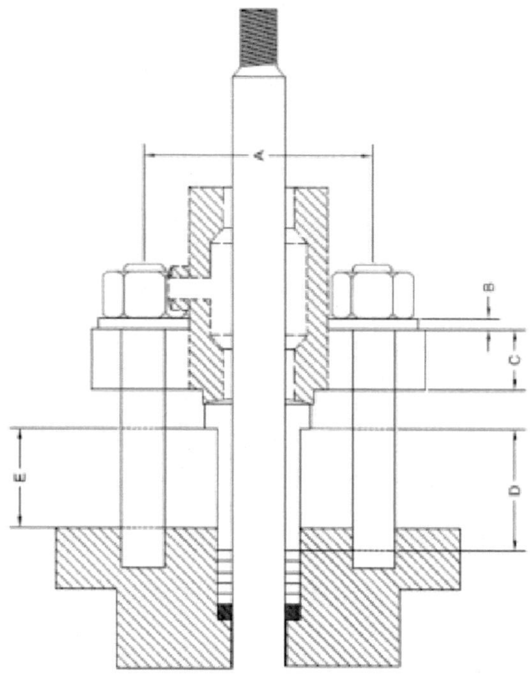

FIGURE 4.5 Bonnet test fixture as per API 622.

Table 4.2 Bonnet test fixture dimensions as per API 622.

Item	1/8 in. Packing fixture	1/4 in. Packing fixture
A	48.79 mm (1′.92 in.)	1′01.60 mm (4.00 in.)
B	3.1′ 8 mm (0.125 in.) max.	3.1′8 mm (0.125 in.) max.
C	25.4 mm (1′.00 in.)	38.2 mm (1′.50 in.)
D	50.8 mm (2.00 in.)	50.8 mm (2.00 in.)
E	Gland height shall be measured and recorded at the beginning and end of test	Gland height shall be measured and recorded at the beginning and end of test

Table 4.3 Stem/stuffing box dimensions as per API 622.

Item	1/8 in. Packing fixture	1/4 in. Packing fixture
Stem diameter	11.05–11.1 mm (0.434–0.437 in.)	25.2–25.4 mm (0.992–1.000 in.)
Stem straightness	Max. 0.04 mm per 300 mm (0.0016 in. per 12 in.)	Max. 0.04 mm per 300 mm (0.0016 in. per 12 in.)
Stem cylindricity	0.04 mm max. (0.0016 in. max.)	0.04 mm max. (0.0016 in. max.)
Stem surface finish	0.40–0.80 μm Ra (16–32 μ-in. Ra)	0.40–0.80 μm Ra (16 p-in. Ra to 32 μ-in. Ra)
Stuffing box diameter	17.46 mm + 0.06 mm or − 0.0 mm (0.688–0.690 in.)	38.1 mm + 0.25 mm or − 0.0 mm (1.500–1.510 in.)
Stuffing box depth	19.05 mm ± 1.57 mm (0.75 in. ± 0.062 in.)	44.5 mm ± 1.57 mm (1.75 in. ± 0.062 in.)
Stuffing box surface finish	3.20 μm Ra + 1.25 μm Ra or - 0.625 μm (125 μ-in. Ra + 50 μ-in. Ra or - 25 μ-in.)	3.20 μm Ra + 1.25 μm Ra or − 0.625 μm (125 μ-in. Ra + 50 μ-in. Ra or − 25 μ-in.)
Gland bottom machined flat	0.15 mm (0.006 in.) max.	0.15 mm (0.006 in.) max.
Gland to stuffing box diametrical clearance	0. 06–0.19 mm (0.0025–0.0075 in.)	0.13–0.38 mm (0.005–0.015 in.)
Stem to gland (flange) diametrical clearance	0.25–0.38 mm (0.010–0.015 in.)	0.5–0.8 mm (0.020–0.030 in.)
Gland stud diameter	3/8 in.—16 UNC (2 pieces)	5/8 in.—11 UNC (2 pieces)

Fig. 4.6 illustrates the test fixtures, including the actuator, stem, gland, and test chamber, filled with methane and insulated to maintain the required temperature. The test fixture could be prepared in simulation of a valve in pressure class 300 equal to 50 bar and 2″ size for 1/8″ packing size, or pressure class 300 equal to 50 bar and 4″ size

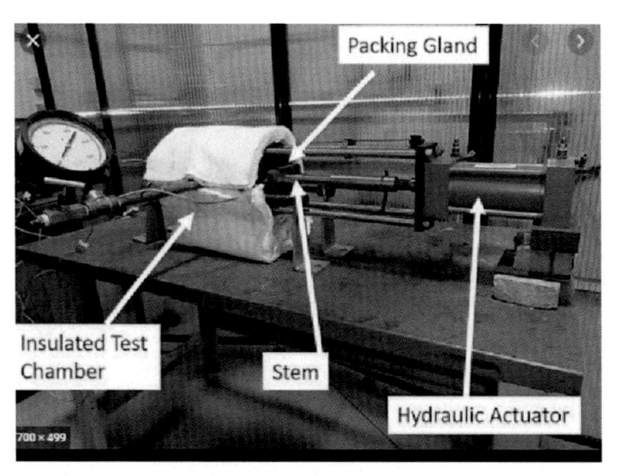

FIGURE 4.6 Test fixture as per the API 622 standard. *Courtesy: Pumps and Systems.*

for 1/4″ packing size. However, it is common in the industry to standardize the test rig to simulate a 4″ wedge gate valve in pressure class 300.

4.2.1.1.1.2 API 622 test fluid The test fluid in API 622 shall be a dry methane gas with 97% minimum purity subject to a temperature range from ambient to 260°C (500°F) and a pressure from 0 psi to 600 psi. Appropriate safety precautions should be taken in the test with methane, as it is a flammable fluid.

4.2.1.1.1.3 API 622 mechanical and thermal cycles Packing in the test rigs shall be subjected to 1510 cycles and 5 thermal cycles for 6 days. Mechanical and thermal cycles begin with the test fixture at ambient temperature. Fig. 4.7, extracted from API 622, illustrates the details of the thermal and mechanical cycles.

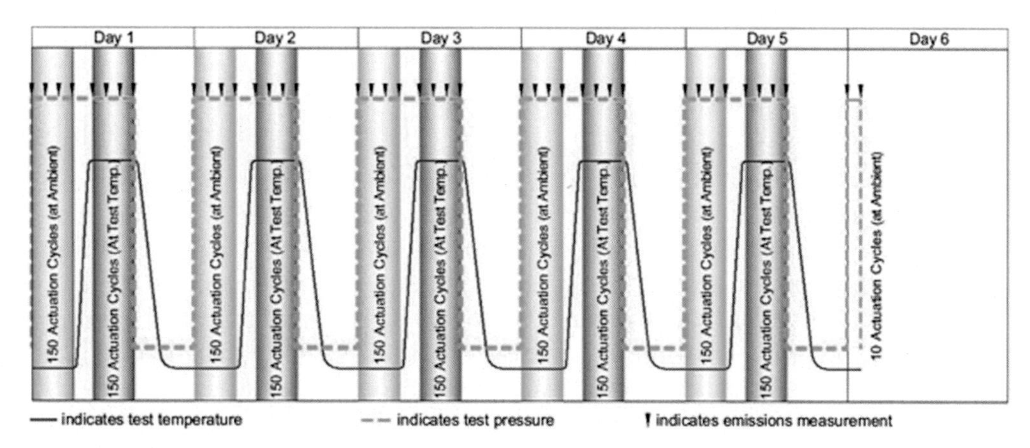

FIGURE 4.7 Mechanical and thermal cycles as per API 622.

The figure shows 300 mechanical cycles per day during the first 5 days, followed by 10 mechanical cycles on day 6. On the first 5 days, there are 300 mechanical cycles on each day; half of the mechanical cycles on each day, equal to 150 mechanical cycles, are combined with a thermal cycle. There is one thermal cycle per day on each of the first 5 days, and no thermal cycle on day 6, which is the last day.

The pressure could be changed temporarily during heating and cooling, but the cycling pressure shall be maintained at 4137 kPag ±34 kPa, which is equal to 600 psig ± 5psig. Mechanical cycles, shown in blue in Fig. 4.7, which are 150 cycles per day during the first 5 days, are done at ambient temperature, which is between 15°C (59°F) to 40°C (104°F). The 150 mechanical cycles per day during the first 5 days are performed at 260°C ± 3°C (500°F ± 5°F).

Two points related to the test procedure could spark discussion: The first point is the duration of the test, which at 6 days is quite long. The second discussion is related to the test pressure, which is standardized to 600 psig on average, which is around 41.4 bar. The test pressure provided in the standard partially covers ASME pressure class 300. However, the test pressure seems too low for valves that are designed in higher ASME pressure classes, such as class 600, 900, 1500, and 2500.

4.2.1.1.1.4 API 622 packing selection and installation Test packing should be selected from a production lot supplied by the manufacturer or from distributor stock. The cross section of packing should be either 1/4″ or 1/8″. The packing for the fugitive emission test should be selected as per the valve manufacturers' installation standard. The amount of load or stress that is applied on the packing through the gland flange stud and nut should not exceed 25,000 psi as per the API 622 standard. Some of the main instructions for packing installation are provided as follows, as per the standard:

- Components related to packing installation, such as the gland, gland flange, and packing, should be inspected prior to installation to make sure that they are free from damage.
- No contact is tolerated between the gland and stem.
- The height of the gland flange shall be measured.
- Fasteners and washers shall be lubricated.

4.2.1.1.1.5 API 622 leak measurement Fugitive emission test leakage monitoring equipment should contain an organic vapor analyzer with an integral data logger or a signal output for data collection related to the emission. The leakage monitoring equipment during the test should be set in flame ionization detection (FID) mode in order to detect carbon and other organic compounds. The minimum and maximum detection levels of emissions during the test are 1.0 parts per million (ppm) and 100 ppm, respectively. The maximum allowable leakage during the test is 100 ppm at the stem and bonnet seal, as per the API 622 standard, with zero adjustment of the stem seal. In addition, the maximum response time in which to measure the amount of leakage is 10 s. It is important to inspect the leakage test device before the test to ensure that the device is working properly and that no fouling has occurred on the detection probe. Also, the test device must be calibrated prior to the test to make sure that the measured leakage is accurate and valid during the test. A methane leak detector probe (sniffer) is connected to the actuator. Fig. 4.8 illustrates the complete fugitive emission test setup as per the API 622 standard. The orientation of the stem is horizontal.

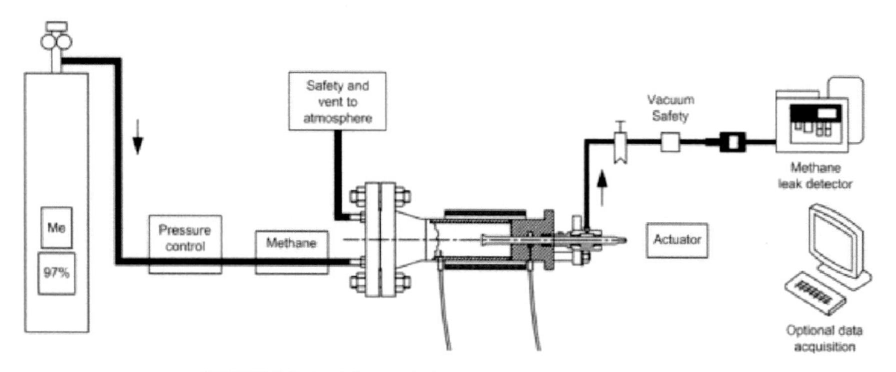

FIGURE 4.8 Fugitive emission test setup as per API 622.

Leak measurement shall be taken at the start of each thermal cycle and after the completion of every 50 mechanical cycles. For each leak measurement, 10 readings should be taken over a 1-min duration. The average reading should be calculated and recorded as the amount of leakage. Any reading above 50% of the average for a leakage rate of more than 10 ppm should be ignored and a new reading should be taken. It is important to measure leakage from the stem and bonnet while the stem is not moving and is in static condition. The connection between the test fixture and the methane leak detector should be through a tube with an internal diameter not larger than 6.3 mm or 1/4″. Table 4.4 provides a packing test result according to the API 622 standard.

Table 4.4 Test results for packing as per API 622.

Test parameter	Results of test 1	Results of test 2
Average test pressure	600 psi (41.4 bar)	602 psi (41.5 bar)
Number of actuation cycles completed	1510	1510
Number of thermal cycles	5	5
Number of packing adjustment	No adjustment	One adjustment after 600 cycles
Average leakage through the test	80 ppm	41 ppm
Maximum leakage throughout	190 ppm	67 ppm

Question: Based on the results of the fugitive emission test as per API 622 given in Table 4.4, are the packings acceptable?

Answer: The accepted packing report as per API 622 should contain three main elements:

1. The fugitive emission test report with a maximum measure leakage from the stem that does not exceed 100 ppm with zero stem seal adjustment. Test result 1 in Table 4.4 shows that the maximum leakage from the stem is 190 ppm, so the packing in test #1 is rejected. Test #2 shows the maximum leakage as 67 ppm, which is less than 100 ppm, but the packing was adjusted once. Therefore, the packing is rejected on the basis of test #2 as well.
2. A completed corrosion test report.
3. A completed material test report.

4.2.1.1.2 API 622 packing corrosion test

A corrosion test is included in the API 622 standard to measure the corrosion of the packing in both hot and cold temperature ranges. In addition, this test can provide some means for evaluating the effect of corrosion inhibitor on the packing and the effect of galvanic corrosion on the valve stem areas in contact with the packing. The test fixture for the packing corrosion test is typically arranged with one ring of 1/4″ packing or two rings of 1/8″ packing. The cold corrosion test is performed at ambient temperature and the hot corrosion test is performed at an average temperature of 149°C equal to 300°F. No acceptance or rejection criteria is provided for the packing corrosion test in API 622.

4.2.1.1.2.1 Ambient temperature corrosion test The ambient temperature corrosion test is performed at ambient temperature in which the packing sample is soaked in demineralized water and then inserted into a small fixture. The packing is compressed around a metallic specimen. In fact, the packing is installed around a test specimen in the shape and material of the metallic ring representing the valve stem material. The steel ring sample should be in martensitic stainless steel grade 410, which contains 13% chromium. The fixtures are stored and kept in an enclosed area in which they are suspended over a water bath. This arrangement ensures a moist environment for 28 days, which is the test duration. The metal sample is then inspected with an X-ray to quantify the percentage of corrosion and the depth of corrosion types such as pitting.

4.2.1.1.2.2 High-temperature corrosion test The high-temperature corrosion test uses a test fixture similar to that of the ambient temperature corrosion test. The exception is that in a high-temperature corrosion test, the temperature of the demineralized water is 149°C equal to 300°F on average instead of ambient temperature. In addition, the pressure of the demineralized water bath is 45 bar on average with a variance of ± 2.25. The duration of this test is 35 days instead 28 days. The same postexposure analysis and measurements are conducted as for the ambient temperature corrosion test.

Fig. 4.9 illustrates the results of the ambient and hot temperature corrosion tests of the metal sample with and without any contact with the packing. The corrosion test result does not show any sign of corrosion on the metal in contact with the packing.

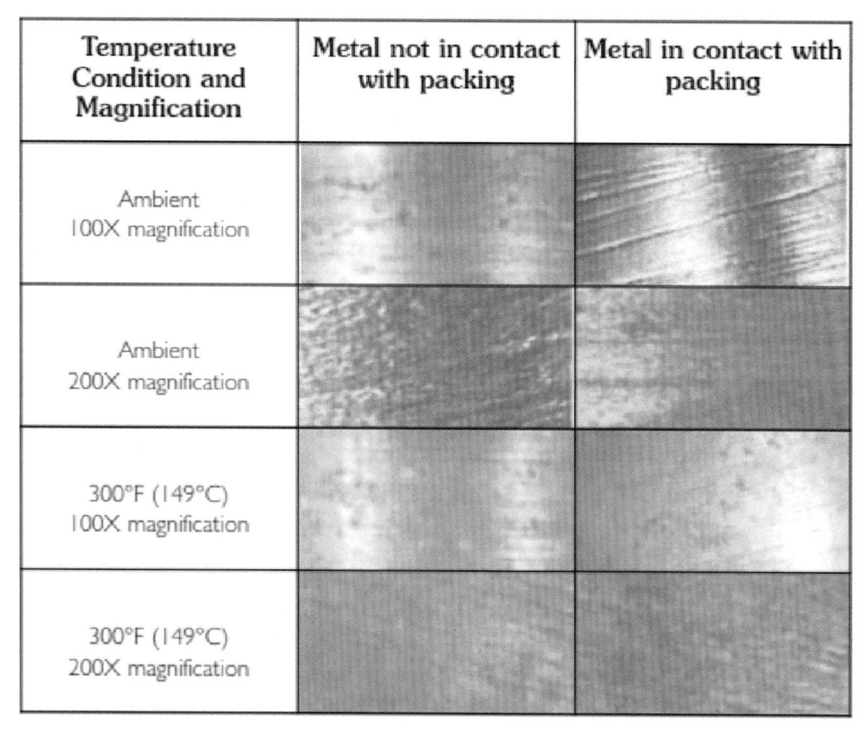

FIGURE 4.9 Corrosion test result as per API 622. *Courtesy: Valve World.*

4.2.1.1.3 API 622 packing material test

The packing material test evaluates weight loss, density, lubricant content, and leachable as per the procedure provided in the API 622 standard. No rejection or acceptance criteria for the packing material test is provided in API 622.

A graphite weight loss test is performed by recording the weight of the packing. The oven is preheated to 150°C and the sample ring is placed in the oven for 1 h. Then the sample is removed from the oven and cooled down to room temperature. Its weight is then recorded. Then the oven temperature is increased to 260°C and the sample is kept in the oven for 1 h; the sample ring is then removed from the oven and the weight is recorded. Then the oven temperature is increased to 538°C and the sample is placed in the oven for 1 h; the sample is then removed from the oven and the weight is measured and recorded.

Density of the packing is determined by dividing the test ring sample weight mass by the sample volume.

Lubricant content is measured through two different sets of tests. The first test is performed to measure the percentage of the polytetrafluoroethylene (PTFE) content. The role of PTFE as a type of lubricant in the valve packing is very important. In fact, PTFE improves both the strength and lubrication capability of the packing. The proposed method to measure the PTFE packing in the standard is to measure the percentage of fluorine content in the packing through one of the ASTM D1179 or ASTM D4327 standards. The results obtained should be divided by 0.76 to determine the approximate amount of PTFE in the packing. The second procedure is given in the standard in order to measure the wet lubricant in the packing.

Leachable compounds: Testing is performed to measure the packing's leachable chloride content, as per ASTM D512, and fluoride content, according to ASTM D1179 or ASTM D4327. Leachable compounds such as chloride and fluoride should be limited in the valve stem packing, which is made in graphite mainly to prevent corrosion and leakage from the packing.

4.2.1.2 API 624 scope

The API 624 standard was published for the first time in 2014. It specifies the requirements and acceptance criteria equal to a maximum of 100 ppm leakage rate for fugitive emission testing of rising and rising–rotating stem valves with packing previously tested in accordance with the API 622 standard. One point of discussion here is about the 100 ppm maximum allowable leakage permission in API 624 as well as the API 622 2018 edition. Historically, the United States in the first LDAR programs targeted valves with a leaking amount of 10,000 ppm. After that, the maximum allowable leakage from the packing was set at 500 ppm, which was reflected in the older revisions of the API 622 standard. Recently, a more restricted limit of 100 ppm is provided for valves in the oil and gas and petrochemical industries. This change of leakage limit to 100 ppm has led many valve manufacturers to change their production process in the last 10 years, and many developments in valve sealings and packing have been implemented in recent years to meet the new leakage limit requirements delineated in the standard.

Packing should be suitable for temperatures between −29°C and 538°C. Specified test in API 624 is required for valves covered by the API 600, 602, 603, and 623 standards, which generally include wedge gate and globe valves and are listed in Table 4.5. API 602 covers small gate, globe, and check valves in size ranges of 4″ and smaller. API 600 covers gate valves in 4″ size and larger. API 603 covers gate valves in size ranges from 1/2″ to 24″ in corrosion-resistant alloys such as stainless steels or nickel alloys. API 623 was released in 2013 to develop a new design for globe valves to prevent cavitation. The leak detection method in API 624 is taken from EPA method 21. It should be noted that gland bolt retightening is not allowed in the type testing as per API 624. Type testing has been defined in Chapter 1 of the book.

Table 4.5 Proposed valve types, sizes, pressure classes, and standards for the fugitive emission test as per API 624.

Type of valve	Size	Pressure class	Standard
Gate and globe	3/4"	800	API 602
Gate and globe	1 1/2″	800	API 602
Gate and globe	3/4″	1500	API 602
Gate and globe	1 1/2″	1500	API 602
Gate	4″	150	API 600
Gate	4″	600	API 600
Gate	4″	1500	API 600
Gate	12″	150	API 600
Gate	12″	600	API 600
Gate	12″	1500	API 600
Gate	20″	600	API 600
Gate	20″	1500	API 600
Gate	3/4″	150	API 603
Gate	1 1/2″	600	API 603
Gate	4″	150	API 603
Gate	4″	600	API 603
Gate	12″	150	API 603
Gate	12″	600	API 603
Gate	20″	150	API 603
Gate	20″	600	API 603
Globe	4″	150	API 623
Globe	4″	600	API 623
Globe	4″	1500	API 623
Globe	12″	150	API 623
Globe	12″	600	API 623
Globe	12″	1500	API 623

Testing certain valves can qualify other types of valves without any need to apply a test according to the following rules:

- For API 602 valves, a nominal pipe size (NPS) 3/4 class 800 test valve may be used to qualify all NPS 1 and smaller valves up to and including class 800. For API 602 valves, an NPS 11/2 class 800 test valve may be used to qualify all valves NPS 11/4 through NPS 21/2 up to and including class 800.
- For API 602 valves, an NPS 3/4 class 1500 test valve may be used to qualify all valves NPS 1 and smaller in class 1500. For API 602 valves, an NPS 11/2 class 1500 test valve may be used to qualify all valves NPS 11/4 through NPS 21/2 in class 1500.
- For all other valves covered by API 600, 603, and 623, NPS 4 qualifies all smaller diameters and one diameter larger and one pressure class lower than the test valve.
- For valves larger than NPS 4, the test valve qualifies valves from two nominal sizes smaller to one nominal size larger and one pressure class lower than the test valve.

4.2.1.2.1 API 624 fugitive emission test

4.2.1.2.1.1 API 624 test fluid The medium used for the fluid test as per API 624 shall be methane with 97% purity, like the test fluid used in API 622. It should be noted that the methane used during the test is a pressurized flammable gas and proper safety measures should be taken.

4.2.1.2.1.2 API 624 mechanical and thermal cycles In the API 624 fugitive emission test, valves are subject to a total of 310 mechanical cycles and 3 thermal cycles as illustrated in Fig. 4.10. API 624 test lasts for almost 4 days. For the first 3 days, 50 mechanical cycles are conducted at ambient temperature and 50 mechanical cycles at an elevated test temperature equal to 260°C/500°F plus and minus 5%. The last or fourth day involves 10 mechanical cycles at ambient temperature. The test pressure should be either 600 psi or the valve design pressure class as per ASME B16.34, whichever is lower. ASME B16.34 is an international valve design standard. Again, the point of discussion about the test pressure is whether a valve in a pressure class of 1500 equal to 250 bar should be tested with 600 psi methane pressure which is around 41.4 bar—much lower than the valve design pressure.

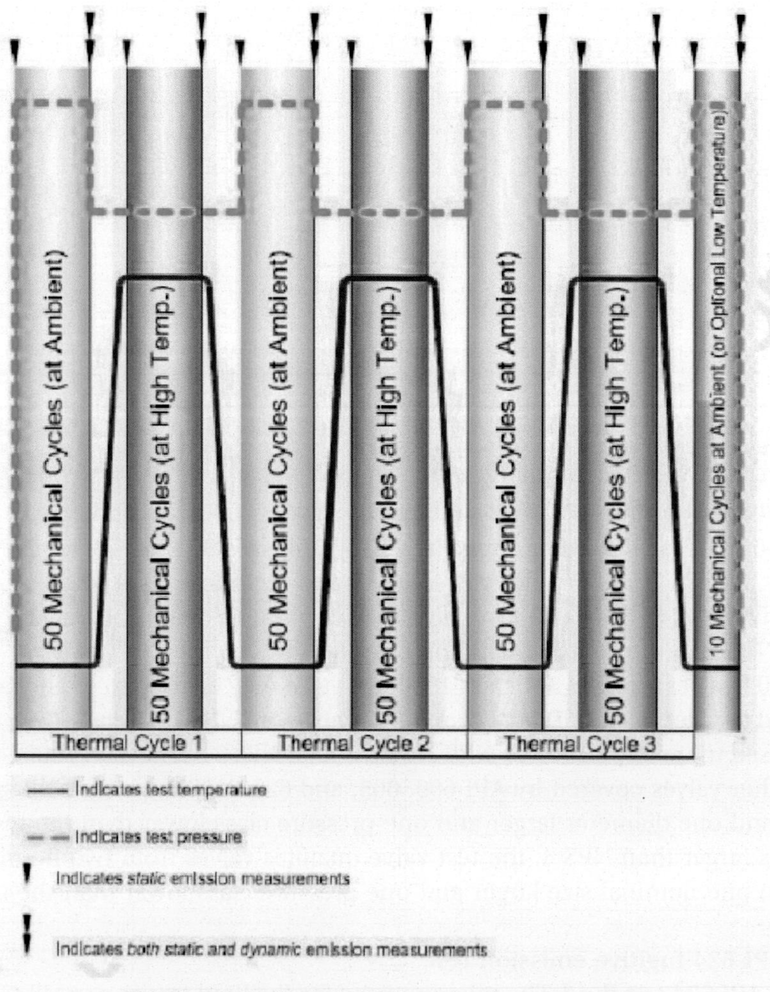

FIGURE 4.10 Mechanical and thermal cycles as per API 624.

The valve is heated through using an external heat source such as a blanket or coil. The test may be performed by using an actuation device on the valve for cycling the valve. The orientation of the valve stem is on the vertical line with the bore horizontal. Two thermocouples as per Fig. 4.11 should be connected to the valve; one shown as TC-1 is connected to the stuffing box, which controls the test temperature. The second, TC-2, is on the external body close to the flow path.

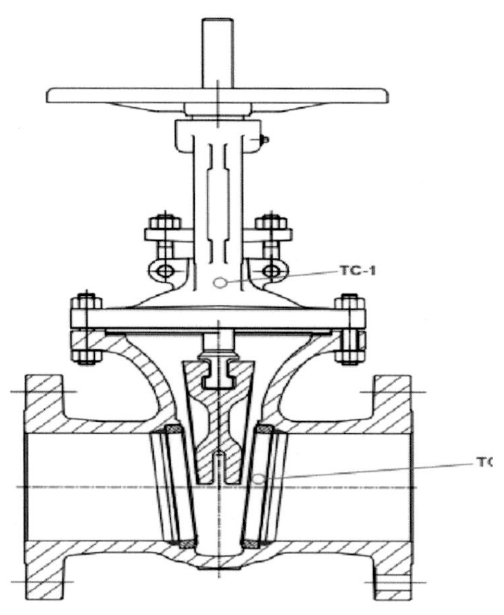

FIGURE 4.11 Thermocouples located on a valve undergoing a fugitive emission test as per API 624.

Fig. 4.12 illustrates the test setup according to API 624 for a 20″ gate valve in pressure class 1500. The valve is manufactured based on the API 600 standard.

FIGURE 4.12 Packing fugitive emission test setup for a 20″ gate valve in pressure class 1500. *Courtesy: Valve World.*

4.2.1.2.1.3 API 624 leak measurement Packing leakage measurement should be performed around the outside diameter of the stem and packing and the highest level of leakage rate should be recorded. It is important to measure leakage during both static and dynamic stem modes as per Fig. 4.10. The leakage measurement method is sniffing using a detection probe. Sniffing leak detection is explained later in this chapter. In addition, leakage from the body and bonnet and auxiliary connections can affect the test result, so they should be corrected prior to the stem leak measurement. Packing adjustment is not allowed in type testing proposed by API 624.

4.2.1.3 API 641 scope

API 641 was published in 2016 to address emissions from quarter turn valves such as ball, plug, and butterfly valves. Fugitive emission testing is more challenging for quarter turn valves due to the possibility of different packing solutions. The testing in API 641 is conducted according to EPA method 21. This standard covers quarter turn valves in sizes less than 24″ and pressure classes lower than 1500. Thus, valves larger than 24″ NPS and valves greater than ASME B16.34 pressure class 1500 are outside the scope of this standard. In addition, valves with a pressure rating at ambient temperature less than 6.89 bar (100 psi) are outside the scope of this standard. One-time stem seal adjustment is allowed for API 641, and if the steam seal adjustment is performed it should be clearly mentioned in the test report. For valves tested according to API 641, the packing material should be qualified in advance through the API 622 standard. If the packing material that is used for valves subjected to a fugitive emission test as per API 641 is out of the scope of API 622, then there is no requirement to test the packing as per API 622 before the valve fugitive emission test. It is not exactly clear which type of packing or stem sealings is not covered by API 622. But the interpretation of this author is that some stem seal materials such as PTFE (Teflon) lip seals and Viton O-rings should be outside the scope of API 622.

4.2.1.3.1 API 641 test fluid

The test medium used for the API 641 test shall be methane with 97% purity, like the test fluid used in API 622 and API 624. It should be noted that the methane used during the test is a pressurized flammable gas and proper safety measures should be taken.

4.2.1.3.2 API 641 mechanical and thermal cycles

A total of 1510 mechanical and 3 thermal cycles should be performed as per API 641. The duration of the test is 4 days (refer to Fig. 4.13). Five hundred mechanical cycles are performed in the first 3 days; 250 mechanical cycles are performed at ambient temperature; and 250 mechanical cycles are performed at high temperature. Ten mechanical cycles are performed on the last day, which is day 4. The tested valve may be equipped with an actuator to provide mechanical cycling of the valve. The value of applied torque on the packing and packing gland should be recorded. The maximum allowable leakage

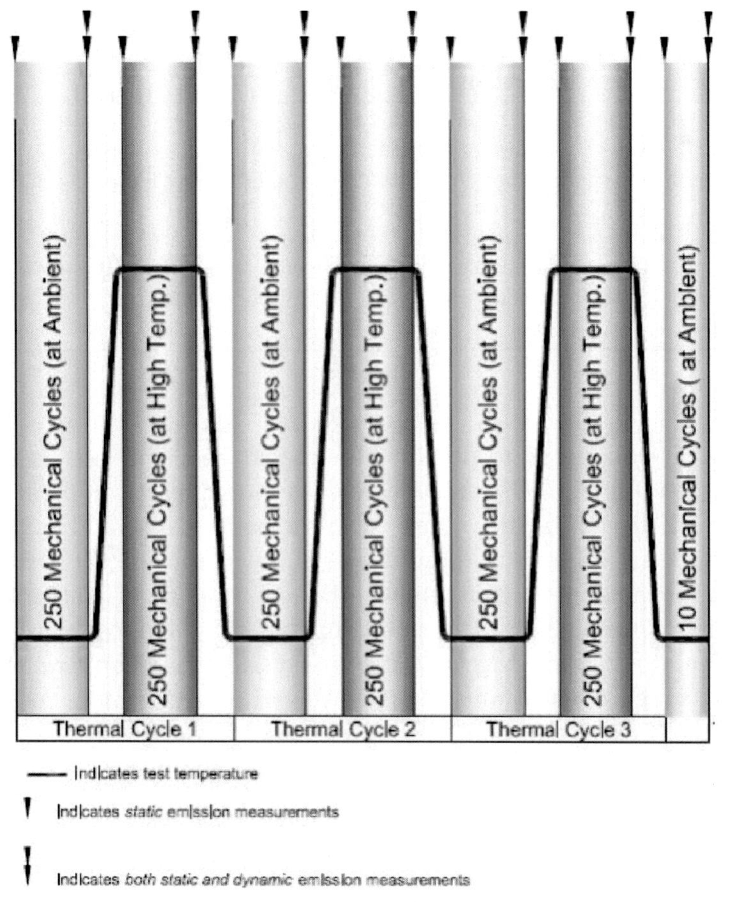

FIGURE 4.13 API 641 mechanical and thermal cycles.

from the packing is 100 ppm, meaning that the test result is given a "pass" indication if the amount of leakage from the valve packing does not exceed 100 ppm. The valve shall be disassembled after the test; its components, including the stem, stem seal, gland, and gland flange, are inspected visually and the conditions of the inspected components are noted in the test report.

4.2.1.3.3 API 641 valve pressure/temperature ratings

Six groups (A, B, C, D, E, and F) are listed in the API 641 standard for the pressure and temperature ratings of the valves. The definition of each group is given in Table 4.6. The

Table 4.6 API 641 valve categories based on pressure and temperature rating.

Valve group	Valve pressure and temperature
A	Valve pressure rating at 260°C (500°F) ≥ 41.1 bar (600 psig)
B	6.89 bar (100psig) ≤ valve pressure rating at 260°C (500°F) < 41.1 bar (600 psig)
C	Valve with a temperature rating ≥ 260°C (500°F) does not comply with the requirements of groups A or B
D	Valve pressure rating as its maximum rated temperature is ≥ 41.1 bar (600 psig)
E	6.89 bar (100 psig) ≤ valve pressure rating at its maximum rated temperature is < 41.1 bar (600 psi)
F	Valve with a temperature rating < 260 °C (500°F) and does not comply with the requirements of groups of D and E

intention of valve grouping is to obtain the elevated and ambient test pressure and temperature as per Table 4.7. P_a and P_e are valve pressures at ambient and elevated temperature, respectively. T_a and T_e are valve temperatures at ambient and elevated temperature, respectively. In fact, the test pressure while at the elevated temperature should be equal to variable P_e. Additionally, the test pressure at ambient temperature should be equal to variable P_a. Ambient temperature is defined in Chapter 1. This grouping and categorization leads to different pressure and temperature test values during the valve's fugitive emission test as per API 641 and adds more complexity to the test procedure and implementation.

Table 4.7 Test pressure and temperature values for each valve category based on API 641.

Valve group	P_a	T_a	P_e	T_e
A	41.1 bar (600 psi)	Ambient temperature	600 psi	260°C
B	Valve pressure rating at ambient temperature or 40 bar (600 psi) whichever is less	Ambient temperature	Valve pressure rating at 260°C	260°C
C	Valve pressure rating at ambient temperature or 41.1 bar (600 psi) whichever is less	Ambient temperature	7 bar (100 psi)	Maximum temperature rating of the valve at 7 bar (100 psi) or 260°C, whichever is lower
D	41.1 bar (600 psi)	Ambient temperature	40 bar (600 psi)	Maximum temperature of the valve
E	Valve pressure rating at ambient temperature or 41.1 bar (600 psi), whichever is less	Ambient temperature	Valve pressure rating at T_e	Maximum temperature of the valve
F	Valve pressure rating at ambient temperature or 41.1 bar (600 psi), whichever is less	Ambient temperature	7 bar (100 psi)	Maximum temperature rating of the valve at 7 bar (100 psi)

Valves with the same quarter turn design as the tested valves could be deemed qualified subject to the following conditions:

- The value for T_e as determined by Table 4.7 is not greater than the value of T_e for the tested valve;
- The value for P_e as determined by Table 4.7 is not greater than the value of P_e for the tested valve;
- The value for P_a as determined by Table 4.7 is not greater than the value of P_a for the tested valve.

It should be noted that any changes in valve stem sealing, such as material, arrangement, and manufacturer, will change the design of the valve and necessitate a new qualification. Change in the obturator type and/or support for the obturator is also considered a new design and requires new qualification. Additionally, any change in the tolerances of manufacturing, surface finishes, and the roughness of the packing and stem requires new qualification.

4.2.1.3.4 API 641 leak test equipment and measurement

Leak test equipment, calibration, and measurement for API 641 are similar to API 622. The test equipment should be inspected prior to the test to make sure that the calibration is still valid for the test.

4.2.2 ISO standards

The International Organization for Standardization (ISO) is a standard-setting organization for different industries such as the oil and gas industry. The main ISO standard for addressing the fugitive emission test is ISO 15848, which has two sections. Section 1 covers the fugitive emission qualification tests for new valve designs, and section 2 focuses on fugitive emission tests for valves that have already undergone qualification fugitive emission testing as per section 1 of the standard. More precisely, part 2 of ISO 15848 is called the production acceptance fugitive emission test of the valves, which is a much simpler and shorter standard and test procedure for fugitive emission compared to part 1 of the standard.

4.2.2.1 ISO standard 15848-1

4.2.2.1.1 ISO standard 15848-1: scope, aim, and objective

ISO 15848-1 is the newest standard that can be used across the globe; this standard covers tests for fugitive emissions from all the possible leakage points of the valves, such as the stem seal and the body and bonnet seals intended to be used for VOCs, hazardous air pollutants (HAPs), and hazardous fluids. The requirements of the ISO standard are very specific and detailed, and the implementation of the test as per this standard could take a relatively long period of time, such as 1–2 weeks. The objective of this standard is to classify different designs and constructions of valves to reduce fugitive emissions.

ISO 15848-1 is only applicable if the fugitive emission requirement has been specified for the valves. Fugitive emissions from the end connection joints, vacuum application, the effect of corrosion, and radiation are excluded from the scope of this standard. The second edition of this standard, which was released in 2015, is reviewed in this section.

This standard covers valves used for isolation purposes, as well as control valves. Isolation valves, also called block valves, are those that are used for stopping/starting or turning the fluid service on/off inside the piping service. In fact, this standard covers gate, globe, and quarter turn valves like ball and control valves.

4.2.2.1.2 ISO standard 15848-1 test preparation

The test shall be applied only on the fully assembled valve. The valve for the test should be randomly selected. The valve must be pressure tested according to ISO 5208 in advance. The valve pressure test may make the packing wet. In that case, the wet packing should be replaced. The valve shall be dry and free of lubricant. The test equipment shall be also clean, dry, and free from oil and lubricant. The test temperature is recorded on three locations of the valve, as illustrated in Fig. 4.14. Measurements at locations #1 and 2 are taken to measure the test temperature. The measurement at location #3 is performed to measure the temperature of the valve in the area close to the stem sealing. Measurement at location #4 is optional if it is not possible to measure the temperature at location #1. The temperature at the measuring points should be stabilized prior to performing the test. The temperature values recorded by the three thermocouples should not have more than 5% variation. If the valve is insulated, then it should be clearly mentioned in the test report. The temperature should be stabilized at all the locations before leak measurement. Temperature stabilization on location #3 should last for at least 10 min before the start of the test. Fig. 4.15 illustrates the stabilization temperature chart for all three locations of the thermocouples when the valve is heated up.

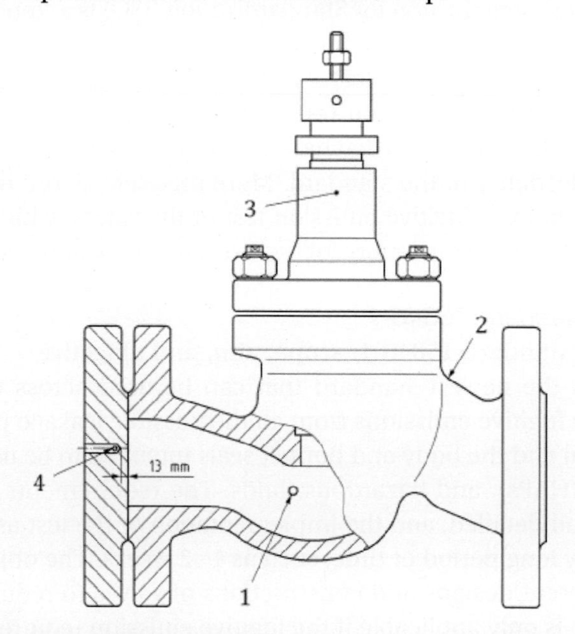

Key

1 location 1: flow path (temperature T_1)
2 location 2: valve body (temperature T_2)
3 location 3: stuffing box (temperature T_3)
4 location 4: optional for flow path (temperature T_1)

FIGURE 4.14 Temperature measurement on the valve as per ISO 15848-1.

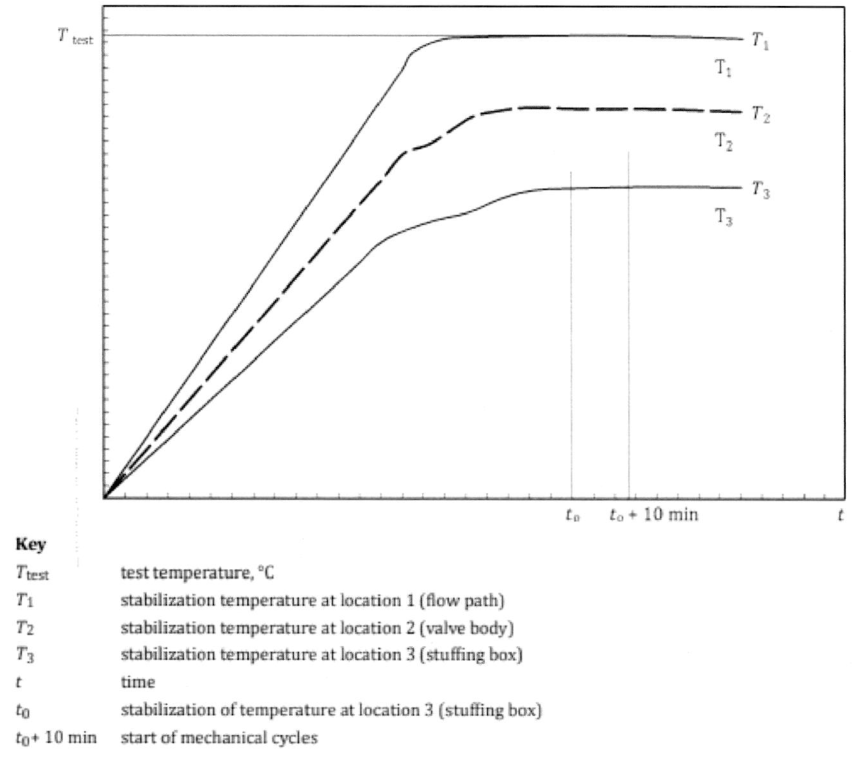

FIGURE 4.15 Temperature stabilization on three valve locations as per ISO 15848.

Fig. 4.16 illustrates the fugitive emission test setup according to the ISO 15848-1 standard for vacuum leak measurement method and helium test fluid. Fig. 4.17 illustrates the fugitive emission test setup according to ISO 15848-1 standard for sniffing leak measurement.

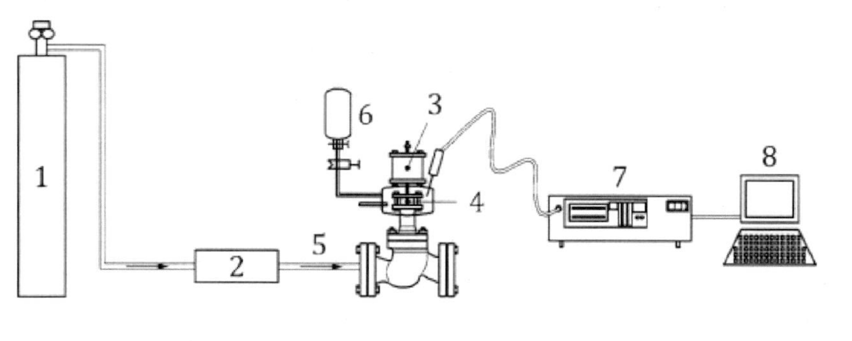

1 helium at 97 % purity
2 pressure control
3 actuator
4 sealed volume or bag
5 helium
6 standard calibrated leak
7 helium mass spectrometer
8 data acquisition

FIGURE 4.16 Test setup as per ISO 15848-1 for helium leak measurement from stem packing through vacuum or bagging leak measurement.

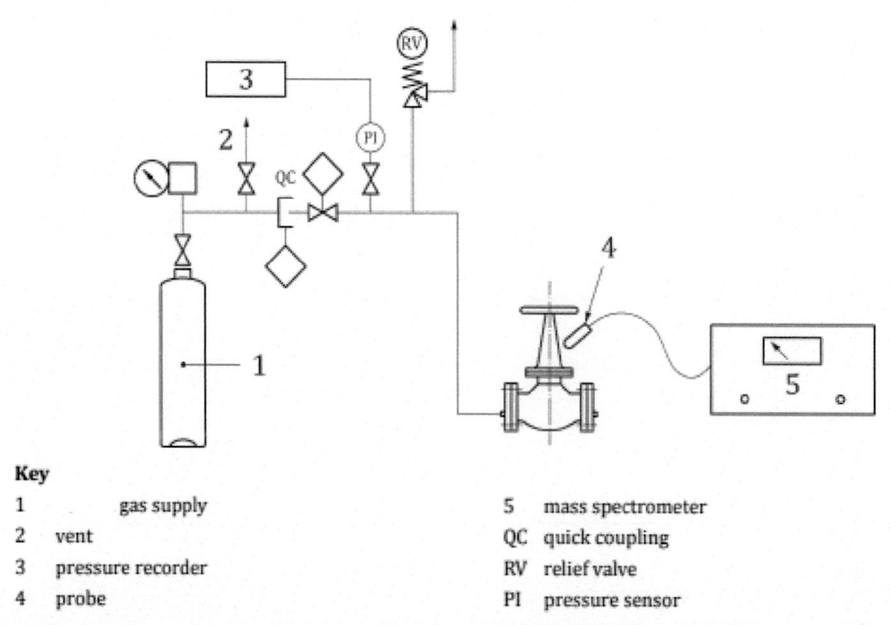

Key

1 gas supply	5 mass spectrometer
2 vent	QC quick coupling
3 pressure recorder	RV relief valve
4 probe	PI pressure sensor

FIGURE 4.17 Test set-up as per ISO 15848-1 for helium or methane leak measurement from the stem packing through sniffing leak measurement. Note 1: Item #1 in the key is the methane gas supply, but it is possible to use helium for the sniffing leak measurement method.

4.2.2.1.3 ISO standard 15848-1 test fluid

The test fluid shall be helium gas of 97% minimum purity or methane of 97% minimum purity. In general, testing with high-pressure gas is potentially hazardous, so safety rules and safety measures should be followed. Methane test fluid is especially flammable, so the test pressure and temperature should be noted and evaluated to prevent possible combustion. The safety measures are not limited to methane; high helium pressure levels or vacuum conditions, especially in conjunction with high temperatures, require strict safety rules.

4.2.2.1.4 ISO standard 15848-1 leakage classes

Leakage class or tightness class is defined as the maximum allowable leakage allowed by the standard. The leakage classes in ISO 15848-1 are different from the helium or methane test fluid for the stem seal. Methane leakage is measured by a detector probe from the stem sealing and body seals. The methane leakage measurement is according to the principles of EPA method 21.

Six tightness classes are defined in the standard for stem seals; three classes for helium and three classes for methane, as illustrated in Table 4.8. Leakage rates for classes starting with “A” are achievable and required for bellow stem seals or stem sealing system for quarter turn valves. Leakage rates for classes starting with “B” are required for PTFE-based packing and elastomeric sealing. Leakage rates for classes starting with “C”

Table 4.8 Allowable leakage rates from the valve stem based on methane and helium fluid as per ISO 15848-1.

Media: Helium	Media: Methane
Leakage rate or class (volumic)	Leakage rate or class
AH $\leq 1.78 \times 10^{-7}$ mbar*L* S^{-1} per mm (millibar liter per second per mm stem diameter)	AM $\leq$ 50 ppm
BH $\leq 1.78 \times 10^{-6}$ mbar*L* S^{-1} per mm (millibar liter per second per mm stem diameter)	BM $\leq$ 100 ppm
CH $\leq 1.78 \times 10^{-4}$ mbar*L* S^{-1} per mm (millibar liter per second per mm stem diameter)	CM $\leq$ 500 ppm

are required and achievable with flexible graphite packing. Therefore, tightness classes starting with "A" are the most stringent leakage classes in this standard. The measured leakage from the body seals for both helium and methane is similar and equal to a maximum of 50 ppm. In general, the vacuum method of leak measurement is used for stem sealing tested with helium. However, sniffing leak measurement is used for measuring leakage from the valve body for both methane and helium tests, as well as leak measurement from the stem during the test with methane fluid. Sniffing and vacuum methods of measuring leakage are explained in detail later in this chapter. The other important point is that there is no correlation between methane and helium leakage measurements. If the stem sealing fails to achieve the required tightness class, the valve may be considered for a lower tightness class.

4.2.2.1.5 ISO standard 15848-1 cycles

4.2.2.1.5.1 Isolation valves Three classes are defined for isolation valves regarding mechanical and thermal cycles, which are shown as CO_1, CO_2, and CO_3.

CO_1: The minimum number of mechanical cycles in CO_1 is 205 with two thermal cycles; a total of 50 cycles at room temperature, 50 cycles at test temperature, 50 cycles at room temperature, 50 cycles at test temperature, and 5 cycles at room temperature, as illustrated in Fig. 4.18. The test temperature is defined based on the temperature classes, which are explained in the next section. If the valve is used for infrequent operation, such as one cycle per month, then applying CO_1 class with the minimum number of cycles given above should be adequate. A maximum of one packing adjustment is accepted for CO_1.

CO_2: An extension of CO_1 that is performed by adding 1295 mechanical cycles and 1 thermal cycle. 795 cycles are done at room temperature followed by 500 cycles at the test temperature. Therefore, the total number of mechanical cycles for CO_2 is 1500 and the total number of thermal cycles is 3. The maximum stem sealing adjustment (SSA) for CO_2 is 2.

CO_3: An extension to CO_2 achieved by adding 1000 mechanical cycles and 1 thermal cycle. Fig. 4.19 illustrates the mechanical and thermal cycles for an isolation valve as per CO_2 and CO_3. Thus, the total number of mechanical cycles as per CO_3 is 2500 and the total number of thermal cycles is 4. The maximum SSA for CO_3 is 3.

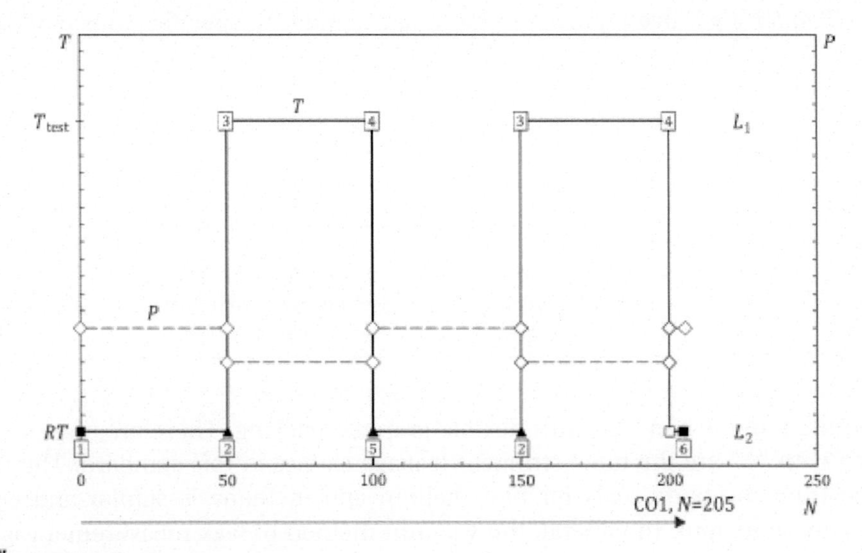

Key

T_{test} test temperature, °C

L_1 measurement of leakage of stem seal

L_2 measurement of leakage of body seal

N number of mechanical cycles

P test fluid pressure

FIGURE 4.18 Mechanical and thermal cycles for an isolation valve (endurance class CO_1) as per ISO 15848-1.

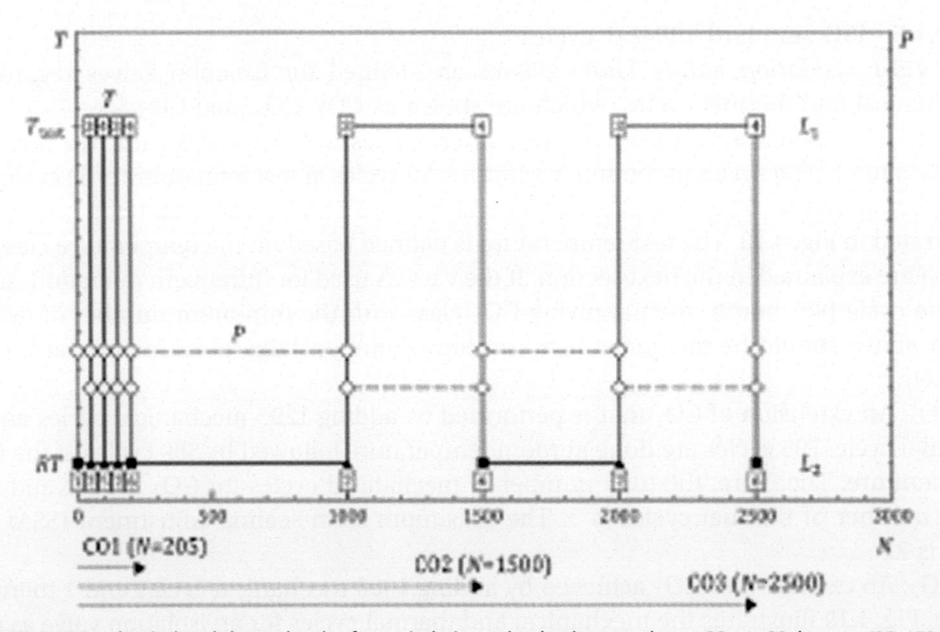

FIGURE 4.19 Mechanical and thermal cycles for an isolation valve (endurance classes CO_2 & CO_3) as per ISO 15848-1.

4.2.2.1.5.2 Control valves CC_1: The minimum number of mechanical cycles for a control valve shall be 20,000 with two thermal cycles. A total of 10,000 mechanical cycles are performed at room temperature and 10,000 mechanical cycles are performed at the test temperature. One packing adjustment is allowed.

CC_2: The minimum number of mechanical cycles for a control valve shall be 60,000 with three thermal cycles. In fact, 40,000 mechanical cycles and 1 thermal cycle is added to CC_1. A total of 20,000 additional cycles are performed at room temperature, followed by 20,000 cycles performed at test temperature. Two packing adjustments are allowed.

CC_3: The minimum number of mechanical cycles for a control valve shall be 100,000 with four thermal cycles. In fact, 40,000 mechanical cycles and 1 thermal cycle are added to CC_2, same as the previous extension added to CC_1. Three packing adjustments are allowed.

All of the cycle classes for the control valves explained above are illustrated in Fig. 4.20, extracted from ISO 15858-1.

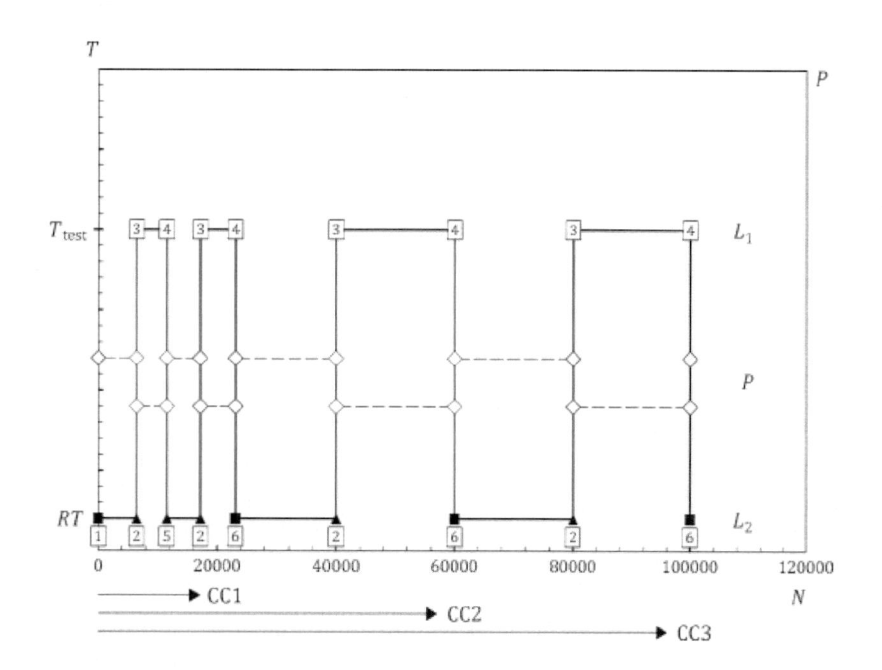

Key

T_{test} test temperature, °C

L_1 measurement of leakage of stem seal

L_2 measurement of leakage of body seal

N number of mechanical cycles

P test fluid pressure

FIGURE 4.20 Mechanical and thermal cycles for a control valve (endurance class CC_1, CC_2, CC_3).

4.2.2.1.6 ISO standard 15848-1 temperature classes

Five temperature classes are defined in Table 4.9 as per ISO 15848-1.

Table 4.9 Temperature classes.

(t-196°C)	(t-46°C)	(tRT)	(t200°C)	(t400°C)
−196°C	−46°C	Room temperature, °C	200°C	400°C

If the test temperature is specified in temperature values other than those mentioned in the table and above zero, the valve should be tested at a temperature class that is the next lower than the specified temperature. As an example, if the test temperature is specified as 215°C, then the valve should be tested at 200°C. If the test temperature would be less than zero and outside the temperature range given in the table, the valve should be tested at the next higher temperature class. As an example, if the test temperature would be −50°C, then the valve should be tested at −46°C. There are some rules for the valve's qualifications according to the testing temperature as follows:

- Testing at −196°C qualifies the valve in the range of −196°C to room temperature;
- Testing at −46°C qualifies the valve in the range of −46°C to room temperature;
- Testing at room temperature qualifies the valve in the range of −29°C to 46°C;
- Testing at 200°C qualifies the valve in the range of room temperature to 200°C;
- Testing at 400°C qualifies the valve in the range of room temperature to 400°C.

To qualify a valve in the range of −46°C to 200°C, two tests are necessary:

- The first test at −46°C qualifies the valve in the range of −46°C to room temperature;
- The second test at 200°C qualifies the valve in the range of room temperature to 200°C.

Note: The valve test pressure should be according to the pressure nominal (PN) or ASME pressure class of the valve at the room temperature and the test temperature. As an example, the body of a valve is in ASTM A216 WCB and CL300 and the valve design temperature is from −29°C to 220°C. Table 4.10 is extracted from ASME B16.34, the design standard for valves, and indicates the pressure and temperature rating for ASTM A216 WCB. The first fugitive emission test on the valve should be performed at a temperature between −29°C and room temperature under a pressure of 51.1 bar, corresponding to the valve pressure rating at both −29°C and ambient temperature. The second fugitive emission test should be performed between room temperature and 200°C. The pressure at room temperature for fugitive emission test is 51.1 bar and the pressure at 200°C is equal to 43.8 bar as per ASME B16.34 standard.

Table 4.10 Pressure/temperature ratings for A216 WCB as per ASME B16.34

Temperature, °C	Working pressures by class, bar						
	150	**300**	**600**	**900**	**1500**	**2500**	**4500**
−29 to 38	19.6	51.1	102.1	153.2	255.3	425.5	765.9
50	19.2	50.1	100.2	150.4	250.6	417.7	751.9
100	17.7	46.6	93.2	139.8	233.0	388.3	699.0
150	15.8	45.1	90.2	135.2	225.4	375.6	676.1
200	13.8	43.8	87.6	131.4	219.0	365.0	657.0
250	12.1	41.9	83.9	125.8	209.7	349.5	629.1
300	10.2	39.8	79.6	119.5	199.1	331.8	597.3
325	9.3	38.7	77.4	116.1	193.6	322.6	580.7
350	8.4	37.6	75.1	112.7	187.8	313.0	563.5
375	7.4	36.4	72.7	109.1	181.8	303.1	545.5
400	6.5	34.7	69.4	104.2	173.6	289.3	520.8
425	5.5	28.8	57.5	86.3	143.8	239.7	431.5
450	4.6	23.0	46.0	69.0	115.0	191.7	345.1
475	3.7	17.4	34.9	52.3	87.2	145.3	261.5
500	2.8	11.8	23.5	35.3	58.8	97.9	176.3
538	1.4	5.9	11.8	17.7	29.5	49.2	88.6

4.2.2.1.7 ISO standard 15848-1 class tightness and marking

Question: A ball valve in 6″ and CL 600 is cycled in the plant at maximum once per month. The temperature of the valve is between −46°C and 200°C. The stem seals are O-rings made of Viton. How should the fugitive emission test be performed on this valve and how should the result be marked on the valve as per the ISO 15848-1 requirements?

Answer: The ball valve is an isolation valve and it is not cycled frequently, so the minimum number of cycles is sufficient for this valve during the fugitive emission test, which is CO_1. The fluid test could be either methane or helium. In addition, since the stem seals are elastomeric O-rings, then the tightness class is "B." The allowable SSA is one time for CO_1. For valves that are tested according to ISO 15848-1, the marking "ISO FE" is applied on the valve stands for ISO fugitive emission. Thus, the complete description of the valve test according to the standard is provided as follows:

ISO FE BH (or BM)— CO_1—SSA 1—t (− 46°C, 200°C)—CL 600—ISO 15848-1.

4.2.2.2 ISO standard 15848-2

4.2.2.2.1 ISO standard 15848-2 scope, aim, and objective

ISO 15848-2:2015 specifies test procedures for the evaluation of external leakage of valve stems or shafts, and the body joints of isolating valves and control valves intended for application with volatile air pollutants and hazardous fluids. End connection joints, vacuum application, effects of corrosion, and radiation are excluded from this part of ISO 15848. The production acceptance test is intended for standard production valves where fugitive emission standards are specified. The second part of ISO 15848 is applicable to valves whose design has been successfully qualified and type tested in advance by ISO 15848-1. The produced valves should be successfully tested prior to the fugitive emission test as per ISO 15848-2. The extension of the test should be agreed between the valve manufacturer and purchaser. As an example, a minimum of one valve from each production lot for each valve type, size, design, and pressure class may be fugitive emission production tested. The packing shall be dry during the test and just one stem seal adjustment is allowed before the start of the fugitive emission test.

4.2.2.2.2 ISO standard 15848-2 test fluid and procedure

The test fluid is helium gas with 97% purity. The leakage measurement method is sniffing method. The test pressure is equal to 6 bar unless otherwise specified, and the test temperature is room temperature.

4.2.2.2.2.1 Shaft sealing test measurement The procedure for measuring leakage from the shaft or stem of the valve is given as follows:

A. Keep the valve in half-open position and pressurize the valve with helium at 6 bar. Measure the stem seal leakage using the sniffing method.
B. Then apply five cycles, which means opening and closing the valve five times.
C. Keep the valve in half-open position after the mechanical cycles and measure the leakage from the stem via the sniffing method performed in step "A."
D. If the leakage values exceed the values given in Table 4.11, the test is considered failed.

Table 4.11 Stem packing/sealing tightness class as per ISO 15848-1.

Class	Measured leakage (ppm)	Remarks
A	≤50	Typically required for bellows or stem sealing for quarter turn valves
B	≤100	Typically achieved with PTFE-based packing or elastomers
C	≤200	Typically achieved with graphite-based packing

4.2.2.2.2.2 Body seal test measurement The procedure for measuring the leakage from the body seal of the valves as per ISO 15848-2 is summarized as follows:

A. Open the valve half way and pressurize the valve with 6 bar helium fluid. The leakage from the body seal should be measured after pressure stabilization.
B. If the measured leakage exceeds 50 ppm, then the test is considered failed. A correction measure should be agreed upon between the valve supplier and the purchaser for the failed valves.

4.2.3 TA Luft/VDI 2440

TA Luft standard or national regulation was established by a German national directive in 1964 with the associated test procedure of VDI 2440. The German technical instructions on air quality control in 2002, known as TA Luft, define the fugitive emission standards for more than 50,000 plants in Germany. The TA Luft standard, unlike the ISO and API standards for fugitive emissions, provides a very general test procedure. TA Luft does not specify any mechanical or thermal cycles for testing the fugitive emission of industrial valves and only addresses stem sealing leakage. Lack of data related to the number of cycles is considered one of the main disadvantages of the TA Luft standard. The fugitive emission testing of valves in accordance with TA Luft typically takes only 1 or 2 days. The test fluid is helium and the method of leakage measurement is vacuum. The test temperature is equal to the maximum operating temperature of the valve and the test pressure is equal to the pressure class of the valve.

Guideline VDI 2440 provides an average, allowable gas emission from different types of stem seals or "gasket systems" as per Table 4.12. Historically, the approach and emphasis of TA Luft is to design and select bellows stem seals for valves to minimize fugitive emission as much as possible. However, there are two main challenges associated with using bellows stem sealing for valves; the first is that a bellows stem seal makes the valve more expensive. The second challenge is that using a bellows stem seal for rotary stem valves such as ball valves and valves with high linear stem motion or stroke is not practical.

Table 4.12 Average gas emission (leakage) from a gasket system or valve stem sealing.

Gasket system (gas sealing)	Allowable leakage rate based on average size of sealing (mg/s × mm of seal)
Stuffing box with packing	1.0
Stuffing box with cap leather, O-ring	0.1
Stuffing box with packing, stuffing box with cup leather, O-ring (with "TA Air Certificates" according to VDI 2440, Section 3.3.1.3)	0.01
Metallic bellows, sealed	0.01
Metallic bellows, sealed (with flat gasket possessing a TA Luft certificate according to VDI 2440)	0.001
Stuffing box with packing and sealing medium/suction, metallic bellows, welded on both sides	No emission (technically leakproof)

4.3 Test fluid (helium vs. methane)

Two types of test fluid are common for fugitive emission tests for valves. The first is helium, which is a very permeable and safe gas to use. Helium is a type of inert gas and is the second lightest gas after hydrogen. Some of the physical properties of this gas are that it is odorless, tasteless, and nontoxic. The second gas is methane, which is not a safe gas and not as permeable as helium. It is clear that helium is a more user-friendly test gas compared to methane. In addition, fugitive emission testing with methane is more costly than performing the tests with helium. Thus, many valve manufacturers prefer to use helium for fugitive emission tests. Some engineers who support helium for the fugitive emission test believe that helium is more fugitive than methane, so the test result with helium is more restrictive. Some who support methane as the test fluid believe that methane is exactly similar to the fugitive compounds from the valves and therefore testing with methane is more accurate. Fig. 4.21 shows a comparison between methane (CH_4) and helium (He) as fugitive emission test fluid.

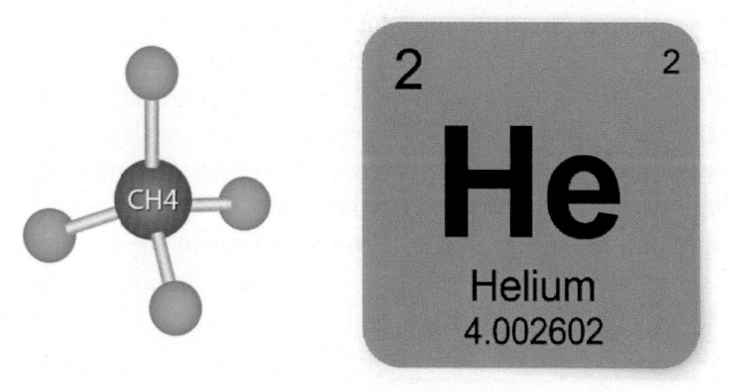

FIGURE 4.21 Methane and helium.

Given that helium is safer and less expensive, the main question is: Why is methane used as a test fluid by some of the major end users in the world as well as in the standards developed in the United States such as API? There are two different reasons. First, historically, many major end users in the oil and gas industry have preferred testing with methane since this gas is more representative of and similar to the VOCs or products leaked from valves during operation in industrial plants. Some of the main end users that prefer fugitive emission testing of the valves with methane are Exxon Mobil, British Petroleum, and Chevron. These companies have developed their own fugitive emission specifications or standards for testing valves based on methane as a test fluid. The second reason for using methane for the tests, especially in the United States, has to do with legislation and legal matters. The EPA in the United States provides a method for measuring leakage from valves and some other components based on parts per million (ppm) using methane as a test fluid. Thus, American end users and US standards use methane as the test fluid for fugitive emission tests.

The second question is: Which standards propose helium for the fugitive emission test? ISO 15848 parts 1 and 2 are based on helium, but an appendix has been added to the ISO 15848 standard to permit methane. Some essential end users like Shell have used the ISO 15848 standard in their MESC and prefer to use helium. Similarly, helium is the preferred fugitive emission test fluid in Germany and among German end users, as using helium is in compliance with TA Luft/VDI 2440. In general, helium test fluid is more common in Europe than the United States.

The other important discussion about test fluid involves the correlation between methane and helium in terms of fugitive amount. There have been a lot of attempts and research conducted to correlate the leakage between these two test fluids—all of which have been unsuccessful. The different physical properties of the gases make correlation impossible. The two gases have different molecule names, Reynolds numbers and different fugitivity at different pressures and temperatures, etc.

4.4 Leak detection methods

Two main approaches are used for leak detection: the vacuum measurement method and the sniffing method, as illustrated in Fig. 4.22. In the vacuum leak detection method, the valve stem, packing, and other top work of the valve is enclosed inside a vacuum area. The vacuum area or tight chamber should be sized to allow the valve actuator to be placed and move for cycling purposes. Helium is injected inside the vacuum enclosure and any leakage from the enclosed area is detected from the spectrometer or leak detector (see Fig. 4.23). Alternatively, sniff leak testing is a basic method of tracing gas detection that uses a sniffer probe to detect the presence of gas and other VOCs on the

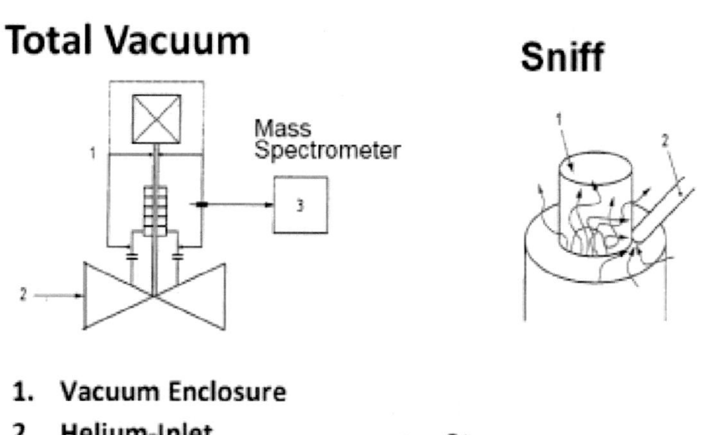

FIGURE 4.22 Sniffing and vacuum leakage detection approaches. *Courtesy: ISO 15848-1*

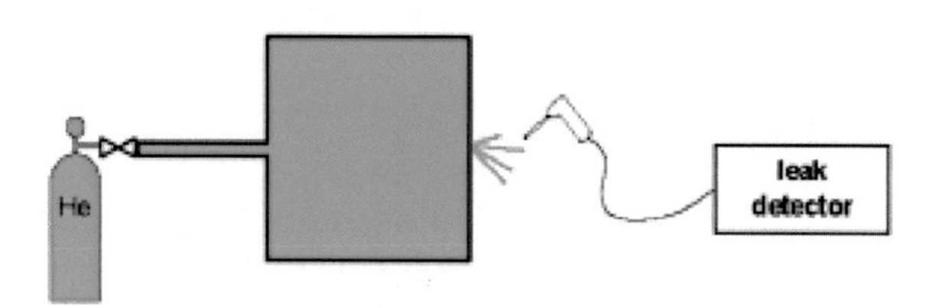

FIGURE 4.23 Vacuum leak detection.

possible area of leakage (in this case the valve packing and stem sealing areas). The figure for vacuum leakage detection is extracted from ISO 15848-1, which is applicable for helium fluid leak detection. The sniffing method of leak detection is also allowed for stem leakage fugitive emission tests performed as per ISO 15848-1 using methane test fluid. In addition, the sniffing method is common for body leak measurement tested with either helium or methane as per ISO 15848-1. Noticeably, the sniffing method of leak detection is more common in the API fugitive emission standards in which methane is used as the test fluid. The method of measuring both stem and body leakage with helium is the sniffing method in ISO 15848-2. It is good to know that leak measurement by sniffing is sensitive to variations in the gaseous atmosphere, which are known as the "weather effect." The atmosphere within the room where the leakage is taking place should be calm and the openings of the room to the environment should remain closed. The bagging method, using the principal of "suck through method," is another method of leak detection for use with helium fluid only as per ISO 15848-1 (illustrated in Fig. 4.24); the bagging method is very similar to the vacuum method.

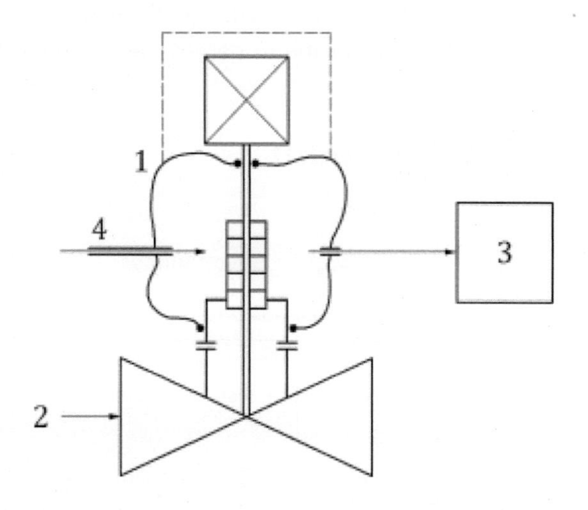

Key

1 bagged volume
2 pressurized helium
3 helium detector
4 balancing tube

FIGURE 4.24 Principal of bagging method (suck through method). *Courtesy: ISO 15848-1.*

4.5 Reasons for test failure

A variety of events experienced during the performance of fugitive emission tests on valves may lead to test failure. Some of the main reasons for test failure are listed in this section.

1. Leakage from the valve stem due to improper stem machining and excessive roughness in the stem areas in contact with the packing. In fact, a higher amount of graphite packing is deposited on the valve stem contact area when the stem is more rough. Graphite deposition on the stem can reduce the volume of the graphite packing and cause leakage. Fig. 4.25 illustrates graphite deposition on two different roughness values of 2.6 μm Ra and 0.2 μm Ra surface finish for a stem under 35 MPa load stress from the packing through the packing gland after a specific number of mechanical and thermal cycles during the fugitive emission test.

FIGURE 4.25 Graphite deposition on the contacted stem area with a roughness of 2.6 μm Ra (left side) and 0.2 μm Ra (right side).

2. Leakage from the valve stem due to excessive loads on the packing from the gland or bushing. Excessive gland stress or load on the packing will increase the deposition of graphite on the stem, which causes leakage.
3. Excessive packing compression under the stem when the valve is installed on the vertical line with the stem horizontal (see Fig. 4.26). This kind of packing leakage is more likely when the packing is tested as per the API 622 standard.

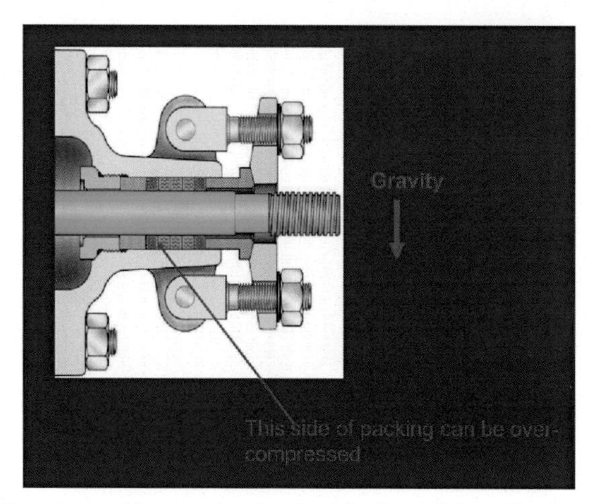

FIGURE 4.26 Excessive packing compression for a horizontal stem. *Courtesy: Valve World.*

4. Packing leakage due to poor material selection, improper arrangement, or size selection including internal and external diameters.

5. Damage to the stem nut, especially its threads, as a result of frequent cycles during the fugitive emission test. Fig. 4.27 illustrates and compares the original bushing on the left with the worn and failed bushing after the fugitive emission test on the right. All the internal threads of the bushing have been removed due to the cycles applied during the fugitive emission test.

FIGURE 4.27 Worn bushing caused by the fugitive emission test on the right compared to the original bushing on the left. *Courtesy: Valve World.*

6. Dripping of lubricant: Lubricants are used on the stem threads and gland flange bolting to prevent galling and friction (see Fig. 4.28). Dripping the lubricant inside the packing box could cause leakage from the packing during the fugitive emission test or during operation. Dripping of the lubricant inside the packing box is not an issue for valves installed on the vertical line with a horizontal stem direction.

FIGURE 4.28 Lubrication of stem threads. *Courtesy: Valve World.*

4.6 Fugitive emission and valve design/manufacturing

The main question answered in this section is: How should valves with low emission that are subject to fugitive emission testing differ from standard valves in design and manufacturing? The simple answer is that the packing, and all the components in contact with the packing, should be evaluated properly to make sure they comply with a low emission requirement. Some of the main parameters in the design and manufacturing of low emission valves are listed as follows:

- The design of the stem, especially adequate roughness on the surface finish, is very important. If the stem surface finish has excessive roughness, a large amount of packing can be deposited on the stem and leakage can occur. On the other hand, an overly smooth finish on the stem would lead to slip-stick phenomena that is defined in Chapter 1.
- The amount of load from the gland flange and gland on the packing should be sufficient. Too low an amount of load may be not enough for the packing to provide a sealing. A high amount of packing load can cause extra friction between the packing and the stem and cause leakage as well as extra torque or load for the valve operation. But in general, a higher amount of load from the gland bolting to the packing is required in low emission applications compared to the standard design for valves. Thus, higher strength gland bolting and gland material are typically required for low emission valves.
- Smooth finish of the gland flange is more important for valves with low fugitive emission applications than for standard valves. As an example, Fig. 4.29 illustrates a gland flange with relatively high roughness which is not suitable for a low emission application.

FIGURE 4.29 A gland flange with high roughness. *Courtesy: Valve World America.*

- Correct dimensions of the stuffing box as well as correct tolerances to keep the packings in place. Stuffing box clearance around the packing can cause extrusion of the packing. Extrusion of the packing material could jeopardize the stem sealing performance and lead to fugitive emission. Therefore, the recommendation of the experts in fugitive emission for valves is to keep the minimum clearance between the packing and packing box, which is also called a stuffing box.

- Packing installation procedure is extremely important for valves with low emission requirements. The proper packing installation procedure should be prepared by the valve manufacturer and evaluated by the purchaser and end user of the valve.

In conclusion, three approaches should be considered together for valve fugitive emission control and prevention. These three approaches are ***design***, ***manufacturing***, and ***testing*** of the valves. ***Design*** includes ***stem surface finish***, ***tolerances on the stem, packing and stuffing box (diameter, surface finish, etc.)***, ***packing and gasket material, and type selection***. ***Manufacturing*** includes ***torque control on the gland and gland bolts*** as well as ***assembly procedure for packing and other components***. ***Testing*** is the third element that is explained in detail in this chapter.

4.7 Conclusions on fugitive emission standards

API 622, 624, and 641 and ISO 15848-1 and 2 are the main international standards for valve fugitive emission tests. Table 4.13 summarizes and compares most of the important test parameters and variations for these five standards.

Table 4.13 Summarization and comparison of the main standards for valve fugitive emission.

	API 622	API 624	API 641	ISO 15848-1	ISO 15848-2
Current revision	3rd edition 2018	1st edition 2014	1st edition 2016	2nd edition 2015	2nd edition 2015
Qualifies	Packing	Valve design	Valve design	Valve design	Produced valve
EPA method 21	Yes	Yes	Yes	Yes	Yes
Prerequisite	None	API 622	API 622	None	ISO 15848-1
Test medium	Methane	Methane	Methane	Helium or methane	Helium
Packing tested in	Fixture	Valve	Valve	Valve	Valve
Type of valves	Any type of valve	Rising stem valves like gate and globe	Quarter turn like ball and butterfly valves	Isolation and control valves	Isolation and control valves
Test pressure	0–600 psi	600 psi or rated valve pressure as per ASME B16.34	Variable	Rated valve pressure as per ASME B16.34	6 bar
Test temperature	Ambient and 260°C	Ambient and 260°C	Variable	Variable	Ambient (room temperature)
Mechanical cycles	1510	310	1510	Variable	5
Thermal cycles	5	3	3	Variable	0
Allowable number of packing adjustment	0	0	1	Variable (1,2, or 3)	0
Acceptance criteria	100 ppm	100 ppm	100 ppm	Variable	Variable
Leak detection method	Sniffing	Sniffing	Sniffing	Sniffing or vacuum or bagging	Sniffing

4.8 Questions and answers

1. Why are fugitive emission tests of industrial valves important?
   - **A.** Regulations regarding environmental issues such as the Clean Air Act
   - **B.** End users' concerns regarding valve fugitive emission
   - **C.** New rules, legislation, and environment awareness
   - **D.** All of the above

   Answer: All of the parameters are correct. Option D is the best answer.
2. Which test parameters and features are similar in API 622 and API 624?
   - **A.** The number of cycles and maximum allowable leakage
   - **B.** Test fixtures and the number of thermal cycles
   - **C.** Zero packing adjustment and maximum allowable leakage
   - **D.** Number of mechanical cycles

   Answer: The number of both mechanical and thermal cycles are different in API 622 and 624. There are 1510 mechanical cycles in API 622, whereas the number of mechanical cycles in API 624 is 310. The number of thermal cycles in API 622 is five, whereas the number of thermal cycles in API 624 is three. The maximum allowable leakage from the packing for both API 622 and API 624 are equal to 100 ppm maximum. The test fixtures are also different in API 622 and API 624. The valve stem is tested in horizontal position during the fugitive emission test as per API 622, whereas the valve stem stands vertically during the test according to API 624. Therefore, options A, B, and D are not correct. Option C is correct, since packing adjustment is not allowed in either API 622 or API 624 standards, and the maximum allowable leakage from packing is 100 ppm.
3. Which test parameter introduces more complication in API 641 compared to API 622 and API 624?
   - **A.** Number of cycles
   - **B.** Variation in the test pressure and temperature
   - **C.** More restricted leakage rate
   - **D.** Duration of the test

   Answer: Table 4.14 compares the number of mechanical and thermal cycles and the duration of the test for API 622, API 624, and API 641.

Table 4.14 Comparing the number of cycles and test duration required by the API 622, 624, and API 641 fugitive emission standards.

Standard	Number of mechanical cycles	Number of thermal cycles	Test duration (day)
API 622	1510	5	6
API 624	310	3	4
API 641	1510	3	4

Thus, the number of cycles and the duration of the test do not add any more complication when comparing API 624 to the other two test standards. The maximum allowable leakage from the packing for all three standards is 100 ppm. Therefore, the only remaining option is option B. The test pressure and temperature in API 641 varies depending on the valve pressure rating or class and the temperature. Six different pressure–temperature classes are defined in API 641, which add more complication to API 641 compared to the other two standards. Thus, option B is the correct answer.

4. Which statement is not correct about the methods used for measuring leakage from valves in the API and ISO standards?
   - **A.** The sniffing method of leak measurement is common in API fugitive emission standards.
   - **B.** ISO 15848-1 allows sniffing, vacuum, and bagging leak measurement methods.
   - **C.** The sniffing method is more common for leakage measurement when the test fluid is methane.
   - **D.** No standard allows for the sniffing leakage detection method for helium test fluid.

   Answer: Option A is correct; the sniffing leak detection method is common in API fugitive emission standards. Option B is also correct since ISO 15848 allows the sniffing, vacuum, and bagging leak measurement methods depending on the test fluid and type of seal in terms of shaft or body seal under the fugitive emission test. Option C is also correct since sniffing is more common for methane test fluid. But the sniffing method is applicable for helium fluid as per ISO 15848-2, so option D is not correct.

5. Which parameters are measured during the fugitive emission test as per ISO 15848-1?
   - **A.** Test pressure and temperature
   - **B.** Test pressure, test temperature, leakage rate, duration of valve mechanical cycle, and actuator force
   - **C.** All parameters in option B plus loading on the packing
   - **D.** Test temperature and duration of mechanical cycles

   Answer: Test pressure and temperature are measured based on ISO 15848-1, but they are not the only parameters measured during the test. Likewise, test temperature and mechanical cycles are not the only parameters measured during the test. Thus, options A and D are not correct. Option C provides all of the parameters measured during the fugitive emission test as per ISO 15848-1 and is thus the correct answer.

6. Which sentence is not correct regarding the comparison between ISO 15848-1 and TA Luft?
   - **A.** ISO 15848-1 is a more stringent and comprehensive standard compared to TA Luft.

B. One stem seal adjustment is allowed in TA Luft, but the number of stem seal adjustments is variable as per ISO 15848-1.

C. Unlike ISO 15848-1, there is no data about mechanical and thermal cycles in TA Luft.

D. The test fluid used in TA Luft and ISO 15848-1 tests is totally different.

Answer: Option A is correct, since ISO 15848-1 is more detailed and comprehensive than TA Luft. Option B is also correct, since one stem seal adjustment is allowed for TA Luft, but the number of stem seal adjustments for fugitive emission tests as per ISO 15848-1 could be one, two, or three depending on the cycle classes. Option D is wrong, since helium is used as the test fluid for TA Luft and ISO 15848-1.

7. Which sentence is correct?

A. Qualification of the packing as per API 622 provides a guarantee of low emission from the valve in which the qualified packing is performed.

B. The only difference between ISO 15848-1 and ISO 15848-2 is that part 1 is used to qualify the valve design and part 2 qualifies the produced valve for fugitive emission.

C. It would be better to test a frequently operated control valve under 100,000 mechanical cycles as per ISO 15848-1.

D. Both helium and methane are applicable for both parts of ISO 15848.

Answer: Option A is wrong, since the qualification of packing as per API 622 does not qualify the valve for low emission. Other tests as per API 624 or API 641 should be performed to qualify the design of the valve in which the packing has been used. Option B mentions the difference between the two parts of ISO 15848, but option B is not correct. There are many important differences between ISO 15848-1 and ISO 15848-2 such as test parameters (number of cycles, test fluid, etc.). Option C is correct since the valve should be tested as per CC_3 with the highest number of mechanical cycles due to frequent operation, and the number of cycles would be 100,000. Option D is not correct since helium is the only test fluid for ISO 15848-2.

8. Fig. 4.30 compares test parameters between the three standards of ISO 15848-1, API 624, and API 641. There are three mistakes associated with the test parameters as per ISO 15848-1. Additionally, there is one mistake related to the test parameters for API 641. What are those mistakes?

Answer: For ISO 15848-1:

- Sniffing is allowed for leak measurement in addition to vacuum and bagging methods.
- Mechanical cycles from 205 to 2500 are performed for isolation valves. Control valves require a higher number of cycles from 20,000 to 100,000.
- The maximum test temperature is according to the five temperature classes defined in the ISO 15848-1 standard, so user-defined is not correct.

For API 641:

- The number of mechanical cycles is 1,500, so 620 in the table is not correct.

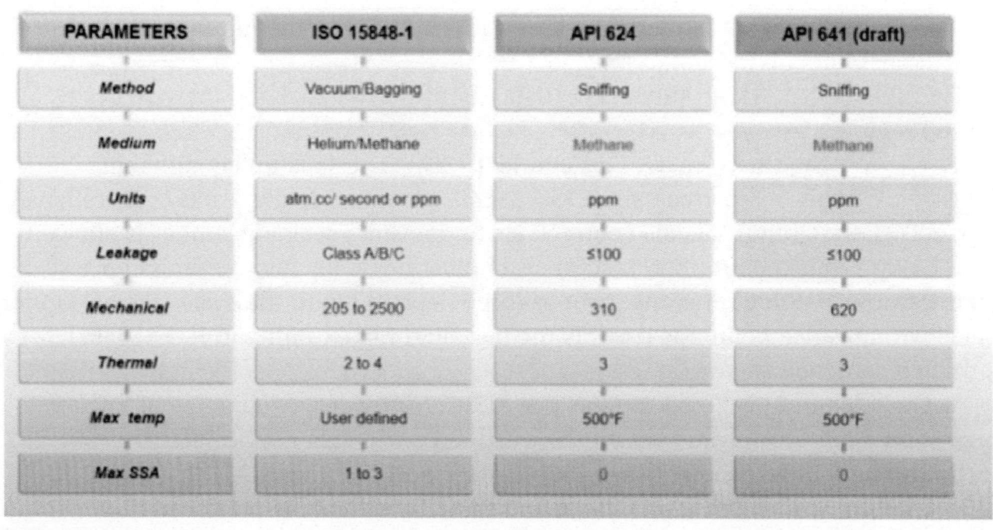

PARAMETERS	ISO 15848-1	API 624	API 641 (draft)
Method	Vacuum/Bagging	Sniffing	Sniffing
Medium	Helium/Methane	Methane	Methane
Units	atm.cc/ second or ppm	ppm	ppm
Leakage	Class A/B/C	≤100	≤100
Mechanical	205 to 2500	310	620
Thermal	2 to 4	3	3
Max temp	User defined	500°F	500°F
Max SSA	1 to 3	0	0

FIGURE 4.30 Test parameters comparison in ISO 15848-1, API 624, and API 641. Note: The table contains three errors.

Bibliography

[1] American Petroleum Institute (API) 600. Steel gate valves—flanged and buttweld ends, bolted bonnet. 13th ed. 2015 [Washington DC, USA].

[2] American Petroleum Institute (API) 602. Gate, globe and check valves for sizes DN 100 (NPS 4) and smaller for petroleum and natural gas industries. 10th ed. 2016 [Washington DC, USA].

[3] American Petroleum Institute (API) 603. Corrosion resistant, bolted bonnet gate valves—flanged and buttweld welding ends. 9th ed. 2018 [Washington DC, USA].

[4] American Petroleum Institute (API) 622. Type testing of process valve packing for fugitive emission. 3rd ed. 2018 [Washington DC, USA].

[5] American Petroleum Institute (API) 623. Steel globe valves—flanged and butt-welding ends, bolted bonnets. 1st ed. 2013 [Washington DC, USA].

[6] American Petroleum Institute (API) 624. Type testing of rising stem valves equipped with graphite packing for fugitive emissions. 1st ed. 2014 [Washington DC, USA].

[7] American Petroleum Institute (API) 641. Type testing of quarter turn valves for fugitive emissions. 1st ed. 2016 [Washington DC, USA].

[8] American Society of Mechanical Engineers (ASME) B16.34. Valves—flanged, threaded, and welding end. 2017. NY.USA.

[9] Boss C, Drago C. Experiences performing API standard 622-type testing of process valve packing for fugitive emissions. Valve World Mag 2009.

[10] Chaney R. Navigating the environment protection agencies (EPAs) fugitive emission standards. MRC Global; 2017 [online] Available from: https://www.mrcglobal.com/Blog/Valves/Fugitive-Emissions. [Accessed 7 November 2020].

[11] International Organization for Standardization (ISO) 15848-1. Industrial valves—measurement, test and qualification—Part 1: classification system and qualification procedure for type testing of valves. 2nd ed. 2015 [Geneva, Switzerland].

[12] International Organization for Standardization (ISO) 15848-2. Industrial valves—measurement, test and qualification—Part 2: production acceptance test of valves. 2nd ed. 2015 [Geneva, Switzerland].

[13] International Organization of Standardization (ISO) 5208. Industrial valve- pressure testing of metallic valves. 4th ed. 2015 [Geneva, Switzerland].

[14] Kirkman B. Discussion—API 624 type testing of rising stem valves equipped with graphite packing for fugitive emissions. Valve World Mag 2014;19(8):78–81.

[15] NELES. Fugitive emission efficiency. 2014 [online] Available from: https://www.neles.com/insights/articles/fugitive-emissions-efficiency/. [Accessed 4 November 2020].

[16] Riedi A. Emission measurements of industrial valves according to TA Luft and ISO 15848-1. Fugitive Emission Control Valve World Mag 2007:51–5.

[17] Valve World America. Ask the expert. 2015 [online] Available from: https://chestertondocs.chesterton.com/Stationary/AskExpertFeb.pdf. [Accessed 11 November 2020].

[18] Venner E. Fugitive emission and control valves. Valve World Mag 2006.

Index

'*Note:* Page numbers followed by "f" indicate figures and "t" indicate tables.'

Printed in the United States
by Baker & Taylor Publisher Services